ELECTRICAL AND TELECOMMUNICATIONS
TECHNICIANS SERIES

EDITORS: A. TRANTER, S. F. SMITH

Electrical principles for technicians

Electrical principles for technicians

Roger Hamilton
Wolverhampton Polytechnic

Oxford   New York   Toronto   Melbourne
OXFORD UNIVERSITY PRESS
1980

Oxford University Press, Walton Street, Oxford OX2 6DP

Oxford London Glasgow
New York Toronto Melbourne Wellington
Kuala Lumpur Singapore Hong Kong Tokyo
Delhi Bombay Calcutta Madras Karachi
Nairobi Dar Es Salaam Cape Town

© Roger Hamilton, 1980

All rights reserved. No part of this publication may be reproduced, stored in a retrieval system, or transmitted, in any form or by any means, electronic, mechanical, photocopying, recording, or otherwise, without the prior permission of Oxford University Press.

British Library Cataloguing in Publication Data

Hamilton, Roger
   Electrical principles for technicians. –
   (Electrical and telecommunications technicians series).
   1. Electric engineering
   I. Title II. Series
   621.3   TK145   79-42794
   ISBN 0 19 859360 0
   ISBN 0 19 859361 9 Pbk

Set by Hope Services, Abingdon
Printed in Great Britain at the University Press, Oxford
by Eric Buckley, Printer to the University

Acc. No.	32101
Class No.	621.3
Date Rec	23 JUN 1981
Order No	F25851

Preface

This book is primarily designed to cover the TEC syllabuses 75/019 and 75/010, Electrical Principles, levels II and III respectively. These units form part of the Certificates in Electronics and Telecommunications. It should also be useful to students studying units U76/359 and U76/360, Electrical and Electronic Principles levels II and III. The main difference is that the latter units contain small amounts of material on semiconductors and a.c. machines which are not covered in this book.

The following should serve as a rough guide to the breakdown of the book into the two levels: Chapter 1 Revision of level I work, Chapter 2 2.1 - 2.3 level II, 2.4 - 2.5 level III, Chapter 3 level II, Chapter 4 level II, Chapter 5 level III, Chapter 6 level II, Chapter 7 level III, Chapter 8 level III, Chapter 9 level III, Chapter 10 mainly II, some III.

Wolverhampton, 1979   R.H.

Contents

1. Introduction — 1
2. D.C. circuits — 8
   - 2.1 Series/parallel circuits — 8
   - 2.2 The principle of superposition — 12
   - 2A Test questions — 16
   - 2.3 Kirchhoff's Laws — 17
   - 2.4 The maximum power theorem — 23
   - 2B Test questions — 24
   - 2.5 Thévenin's theorem — 24
   - 2C Test questions — 31
3. Capacitors and capacitance — 33
   - 3.1 Electric charge and field — 33
   - 3.2 Capacitance — 35
   - 3.3 Capacitors in parallel and in series — 37
   - 3.4 The parallel-plate capacitor — 38
   - 3.5 Energy stored in a capacitor — 39
   - 3.6 Dielectric strength — 40
   - 3.7 Practical capacitors — 42
   - 3A Test questions — 43
4. The magnetic field and electromagnetic induction — 46
   - 4.1 Permanent magnets — 46
   - 4.2 The magnetic effect of an electric current — 47
   - 4.3 Force on a current-carrying conductor — 49
   - 4A Test questions — 52
   - 4.4 Electromagnetic induction (FARADAY) — 54
   - 4B Test questions — 57
   - 4.5 Magnetic circuits — 58
   - 4C Test questions — 69
   - 4.6 Hysteresis — 70
   - 4.7 Self inductance — 71
   - 4.8 Mutual inductance — 73
   - 4D Test questions — 75
5. Inductance and capacitance in d.c. circuits — 76
   - 5.1 L-R circuits — 76
   - 5.2 C-R circuits — 84
   - 5A Test questions — 87

6. A.C. circuits I — 90

    6.1 Alternating quantities — 90
    6.2 Phasors — 95
    6A Test questions — 105
    6.3 Single-phase a.c. circuits — 106
    6B Test questions — 121
    6.4 Series resonance — 122
    6C Test questions — 124

7. A.C. circuits II — 126

    7.1 Power in a.c. circuits — 126
    7A Test questions — 130
    7.2 Q factor — 131
    7B Test questions — 136
    7.3 Parallel circuits — 136
    7C Test questions — 150
    7.4 Parallel resonance — 150
    7D Test questions — 161
    7.5 Three-phase supplies — 162
    7E Test questions — 169

8. Single-phase transformers — 170

    8.1 The perfect transformer — 170
    8.2 Matching — 175
    8.3 Transformer losses — 179
    8A Test questions — 183

9. D.C. machines — 185

    9.1 D.C. generators — 185
    9A Test questions — 190
    9.2 The d.c. motor — 191
    9.3 Shunt motors — 193
    9B Test questions — 198
    9.4 The series motor — 199
    9.5 Motor starters — 202
    9.6 Losses — 202
    9C Test questions — 203

10. Electrical measurements — 204

    10.1 Measurement of current — 204
    10.2 Measurement of voltage — 209
    10.3 Measurement of resistance — 213
    10A Test questions — 217

viii *Contents*

10.4	Measurement of power	221
10.5	The cathode-ray oscilloscope as a measuring instrument	222
10.6	Conclusion	223
10B	Test questions	223

Solutions to test questions   225

Index   229

1 Introduction

This chapter gives a brief summary of the electrical topics which you should have studied in level I and which are needed for level II and III. It is based on the TEC unit 'Physical Science 1' (U75/004). You are advised to attempt all the exercises before proceeding to Chapter 2.

Current, voltage, and resistance

Current, I, is the rate of flow of charge. In a metallic conductor it is carried by a flow of electrons (negatively charged) towards the positive terminal. By convention, current (which was discovered before the electron) is always said to flow from the positive to the negative ends of the conductor.

The unit of current is the ampere (A) which is a flow of charge of one coulomb per second. It flows owing to a potential difference, or voltage, between two points in the circuit. The unit of potential difference is the volt (V).

A continuous current requires a complete circuit which will include a source of electro-motive force (e.m.f.) such as a battery or generator. E.M.F. is also measured in volts.

The opposition which a circuit, or component, offers to a flow of current is called its resistance, R, and is the ratio of voltage to current, i.e. it is the voltage required to produce a current of 1 A in the component. Hence,

(1.1) $$R = \frac{V}{I}.$$

In some components it is found that the current flowing is proportional to the applied voltage. This fact was discovered by Ohm and is known as Ohm's Law. For such a component a graph of voltage against current will be a straight line through the origin, as seen in Fig. 1.1. Fig. 1.2 shows a graph of voltage against current for a component which does not obey Ohm's Law. It is, in fact, the characteristic of a silicon junction diode. The component of Fig. 1.1 is said to be linear, and that of Fig. 1.2 non-linear.

Series and parallel circuits

A number of components are said to be in series if the same current flows through each. Fig. 1.3 shows three resistors in series, with a current I flowing. The diagram also shows the potential differences or voltages across each resistor. It follows from this drawing that the three resistors are equivalent to a single resistor R given by

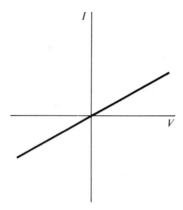

Fig. 1.1. Characteristic of a linear component.

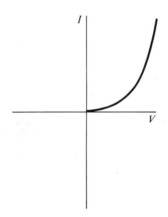

Fig. 1.2. Characteristic of a non-linear component.

(1.2) $$R = R_1 + R_2 + R_3.$$

A number of components are said to be in parallel if the same voltage exists across each. Fig. 1.4 shows three resistors in parallel. Obviously $I = I_1 + I_2 + I_3$ and the three in parallel are equivalent to a single resistor R where

(1.3) $$\frac{1}{R} = \frac{1}{R_1} + \frac{1}{R_2} + \frac{1}{R_3}.$$

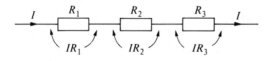

Fig. 1.3. Resistors in series.

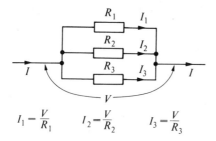

Fig. 1.4. Resistors in parallel.

Note that for two resistors in parallel

(1.4) $$R = \frac{R_1 R_2}{R_1 + R_2}.$$

Resistivity

The resistance of a component depends on the material from which it is made and on its dimensions. Considering Fig. 1.5, the resistance between the ends of the rod of material shown is proportional to the length, l, and inversely proportional to the cross-sectional area, a. Thus the resistance is given by

(1.5) $$R = \frac{\rho l}{a}$$

where ρ is called the resistivity of the material. It is the resistance between opposite faces of a unit cube of the substance. The unit usually used is the ohm-metre.

Temperature coefficient of resistance

Another factor on which the resistance of a component depends is the temperature. Over a small temperature range, resistance varies linearly with temperature. The temperature coefficient of resistance, α, is defined as the ratio of the increase of resistance per degree Centigrade rise of temperature to the resistance at $0°C$. While the resistance of pure metals increases with

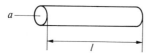

Fig. 1.5.

an increase of temperature, the resistance of some substances, such as carbon, silicon, and germanium, decreases with an increase of temperature. For these substances α is negative.

If the resistance of a material is R_0 at $0°C$ and R_1 at $t_1°C$,

(1.6) $$R_1 = R_0(1 + \alpha t_1).$$

Similarly, if the resistance at $t_2°C$ is R_2

(1.7) $$R_2 = R_0(1 + \alpha t_2)$$

and dividing eqn (1.6) by eqn (1.7) enables us to find R_1 if we know R_2 or vice versa

(1.8) $$\frac{R_1}{R_2} = \frac{1 + \alpha t_1}{1 + \alpha t_2}.$$

Effects of a current

An electric current has three main effects:

(i) A magnetic effect—there is a magnetic field associated with an electric current which will cause a conductor carrying a current to experience a force when situated in a magnetic field. This is the principle of the moving-coil meter and various types of electric motor. Further, if a conductor moves in a magnetic field, an e.m.f. is induced. This is the principle of the electric generator.

(ii) A chemical effect—used, for example, in electroplating.

(iii) A heating effect—obvious from electric fires, fuses, etc.

If a charge Q is moved through a potential V the work done is QV. However, as current is the rate of flow of charge,

$$I = \frac{Q}{t}, \quad \text{where } t = \text{time,}$$

or $Q = It$.

Hence the work done $= QV = ItV$. But power is the rate of doing work and

(1.9) $$P = \frac{\text{work done}}{t} = IV.$$

Hence a current of I (amperes) flowing through a resistance of R (ohms) and producing a voltage drop of V (volts) will develop a power in the resistor (in the form of heat) of IV (watts). Eqn (1.1) applied to eqn (1.9) gives two alternate ways of calculating the power:

(1.10) $$P = I^2 R$$

(1.11) and $$P = \frac{V^2}{R}.$$

Internal resistance of a voltage source

A source of voltage, such as a battery or rotating generator, will normally possess an internal resistance. This is the resistance of the actual battery material or generator coils, and will be given the symbol r. It means that the voltage across the terminals of the source differs from the actual voltage generated (the e.m.f.) by the amount dropped in the internal resistance:

(1.12) $$V = E - Ir,$$

where V = terminal voltage, E = generated e.m.f., and I = current flowing. Obviously, on open circuit, with $I = 0$,

$$V = E,$$

so that E can be measured by using a voltmeter of very high resistance.

In general, if we plot the terminal voltage V of a supply, against the current being delivered, we get a graph such as that shown in Fig. 1.6. r can then be found from

(1.13) $$r = \frac{E - V_1}{I_1}.$$

Multiples and sub-multiples

Table 1.1 shows some common multipliers which are used with physical and electrical units. Note that with the exception of 'kilo' multiples use capital letters and sub-multiples small letters.

TABLE 1.1

kilo	k	10^3	milli	m	10^{-3}
mega	M	10^6	micro	μ	10^{-6}
giga	G	10^9	nano	n	10^{-9}
tera	T	10^{12}	pico	p	10^{-12}

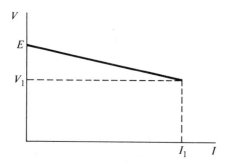

Fig. 1.6.

1A Test questions

1. A voltage of 10 V is applied across a resistance of 5 Ω. Find the current flowing.

2. A current of 3 A flows through a resistance of 2 Ω. Find the voltage across the resistor.

3. A voltage of 2 V is connected across a resistor. If the current flowing is 1 mA find the resistance.

4. Find the voltage required to produce a current of 5 μA in a resistance of 10 MΩ.

5. In each of the four previous questions, find the power being developed.

6. Two resistors of values 20 Ω and 30 Ω are connected in series to a 100 V supply. Find
    (a) the total resistance,
    (b) the current in each resistor,
    (c) the p.d. across each resistor,
    (d) the current flowing from the supply,
    (e) the power developed in each resistor,
    (f) the power taken from the supply.

7. Repeat the previous question if the two resistors are connected in parallel.

8. Three resistors of values 1 kΩ, 2 kΩ, and 3 kΩ are connected in parallel. When the parallel combination is connected to a supply the supply delivers 12 mA. Find
    (a) the voltage of the supply,
    (b) the current in each resistor,
    (c) the power developed in each resistor,
    (d) the total power taken from the supply.

9. An electric kettle has a 2 kW element. Find the resistance of the element if it is designed to operate from a 240 V supply.

10. Calculate the resistance of a 1 km length of copper wire of cross-sectional area 1 mm^2 if the restivity of copper is 1.8×10^{-8} Ωm.

11. Two wires are connected in parallel to a supply and take a total current of 8 A. One wire is made of copper (resistivity 1.8×10^{-8} Ωm) and is 2 m long and of diameter 1 mm. The other wire is made of aluminium (resistivity 2.8×10^{-8} Ωm) and is 1.5 m long. If the copper wire passes a current of 3 A find the diameter of the aluminium wire.

12. A conductor has a resistance at 0°C of 0.6 Ω. If the temperature coefficient of resistance is 0.0043/°C find its resistance at 50°C.

13. A rod of carbon has a resistance at 0°C of 2 Ω and a resistance at 60°C of 1.95 Ω. Find the temperature coefficient of resistance of carbon.

14. A conductor has a resistance of 3 Ω at 20°C. If the temperature coefficient of resistance (at 0°C) is 0.0043/°C find the resistance of the conductor at 80°C.

15. A 100 W, 240 V lamp has a tungsten filament with a resistance at

20°C of 375 Ω. Find the normal operator temperature of the filament if its temperature coefficient of resistance is $45.0 \times 10^{-4}/°C$.

16. A battery has an e.m.f. of 2 V. When it is supplying a current of 2 A its terminal voltage is 1.9 V. Determine the internal resistance of the battery.

17. Two resistors of values 2 Ω and 3 Ω are connected in parallel to a battery of internal resistance 0.2 Ω. If the terminal voltage of the battery is 1.8 V determine
    (a) the e.m.f. of the battery,
    (b) the power developed in each resistor,
    (c) the power developed (wasted) inside the battery.

2 D.C. circuits

This chapter serves as an introduction to circuit theory. Circuit theory deals with methods of 'solving' electrical circuits, that is finding the currents and voltages at various points in a combination of electrical components. Here we shall only deal with direct current circuits, but the various methods used can be applied to alternating current circuits as well.

As we saw in Chapter 1, circuits contain components which may be linear or non-linear, and it is very important when dealing with a circuit theorem or law to understand whether it applies to all circuits or, as is often the case, only to linear ones. For example, Ohm's Law obviously only applies to linear circuits—in fact a linear d.c. circuit can be defined as one which obeys Ohm's Law. Most circuit theorems apply only to linear circuits and, as we shall often be concerned with non-linear circuits such as diodes and transistors, they may seem rather pointless. However, non-linear components are sometimes operated only over a linear part of their characteristics and then the theorems which only apply to linear components can be used. If this is not the case we need to use very different methods of solution, such as the load-line construction which you will meet in your study of electronics.

2.1 Series/parallel circuits

In the previous chapter we saw how to solve simple circuits using the equation

(2.1) $$V = IR$$

together with the equations for finding the total resistance of resistors in series and parallel (eqns (1.2) and (1.3) respectively). It is sometimes more useful when dealing with parallel circuits to use a quantity called the *conductance* of a resistor rather than its resistance.

Conductance, G, is the reciprocal of resistance, and is measured in siemens (S). Thus

(2.2) $$G = \frac{1}{R},$$

(2.3) and $$I = GV.$$

Eqn (1.3) for resistors in parallel can obviously be written

(2.4) $$G = G_1 + G_2 + G_3.$$

The following examples include rather more complicated circuits involving series and parallel resistors.

Examples 2.1

1. Three resistors of $2\,\Omega, 5\,\Omega$, and $3\,\Omega$ respectively are connected in series to a 10 V battery. Find the current flowing and the voltage across each resistor.

Solution

$$\text{Total } R = 2 + 5 + 3 = 10\,\Omega.$$

$$I = \frac{10\,\text{V}}{10\,\Omega} = 1\,\text{A}.$$

Voltage across $2\,\Omega = 1\,\text{A} \times 2\,\Omega = 2\,\text{V}.$

Voltage across $5\,\Omega = 1\,\text{A} \times 5\,\Omega = 5\,\text{V}.$

Voltage across $3\,\Omega = 1\,\text{A} \times 3\,\Omega = 3\,\text{V}.$

Note: $2\,\text{V} + 5\,\text{V} + 3\,\text{V} = 10\,\text{V}$–the voltage of the battery.

2. Repeat example 1 if the battery has an internal resistance of $1\,\Omega$. Also find the terminal voltage of the battery.

Solution

$$\text{Total } R = 2 + 5 + 3 + 1 = 11\,\Omega.$$

$$I = \frac{10\,\text{V}}{11\,\Omega} = 0.91\,\text{A}.$$

Voltage across $2\,\Omega = 0.91\,\text{A} \times 2\,\Omega = 1.82\,\text{V}$

Voltage across $5\,\Omega = 0.91\,\text{A} \times 5\,\Omega = 4.55\,\text{V}$

Voltage across $3\,\Omega = 0.91\,\text{A} \times 3\,\Omega = 2.73\,\text{V}$

Voltage across $1\,\Omega = 0.91\,\text{A} \times 1\,\Omega = 0.91\,\text{V}$

Terminal voltage of battery $= 10 - 0.91 = 9.09\,\text{V}.$

3. Three resistors of $2\,\Omega, 3\,\Omega$, and $6\,\Omega$ are connected in parallel to a 10 V battery. Find the total resistance, the total current, and the current in each resistor.

Solution

$$\frac{1}{R} = \frac{1}{2} + \frac{1}{3} + \frac{1}{6} = \frac{6}{6} = 1$$

$$R = 1\,\Omega.$$

$$\text{Total current} = \frac{10\,\text{V}}{1\,\Omega} = 10\,\text{A}.$$

10 D.C. circuits

Each resistor has a voltage of 10 V across it, hence

Current in 2 Ω = 10 V/2 Ω = 5 A.
Current in 3 Ω = 10 V/3 Ω = 3.33 A.
Current in 6 Ω = 10 V/6 Ω = 1.67 A.

Note: 5 + 3.33 + 1.67 = 10 A—the total current.

Alternate solution

The conductances of the 2 Ω, 3 Ω, and 6 Ω resistors are 0.5 S, 0.333 S, and 0.167 S respectively.

Total conductance G = 0.5 + 0.333 + 0.167 = 1.0 S.

Hence I = 1.0 S × 10 V = 10 A.

Current in 0.5 S   = 0.5 S × 10 V = 5 A.
Current in 0.333 S = 0.333 S × 10 V = 3.33 A
Current in 0.167 S = 0.167 S × 10 V = 1.67 A, as before.

4. Fig. 2.1 shows a circuit containing a battery and four resistors. Find all the currents.

Solution

Resistors of 2 Ω, 3 Ω, and 6 Ω in parallel have a total resistance of 1 Ω (see example 3).

1 Ω in series with 4 Ω = 5 Ω.

Total current = 10 V/5 Ω = 2 A.

Voltage across 4 Ω = 2 A × 4 Ω = 8 V.

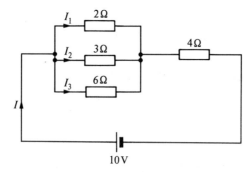

Fig. 2.1.

Hence voltage across parallel combination = 10 − 8 = 2 V. (This may also be found by 2 A × 1 Ω = 2 V.)

Hence $I_1 = 2\,\text{V}/2\,\Omega = 1\,\text{A}$.

$I_2 = 2\,\text{V}/3\,\Omega = 0.67\,\text{A}$.

$I_3 = 2\,\text{V}/6\,\Omega = 0.33\,\text{A}$.

5. The circuit shown in Fig. 2.2 is connected to a battery of e.m.f. 5 V and internal resistance 0.3 Ω. Find the currents shown and the terminal voltage of the battery.

Solution

2 Ω, 3 Ω, and 6 Ω in parallel = 1 Ω (see example 3).

2 Ω and 3 Ω in parallel = $\dfrac{2 \times 3}{2 + 3} = \dfrac{6}{5} = 1.2\,\Omega$.

Total resistance presented to e.m.f. = 1 + 1.2 + 0.3 = 2.5 Ω.

Total current = 5 V/2.5 Ω = 2 A.

Volts dropped in battery = 2 A × 0.3 Ω = 0.6 V.

Terminal voltage = 5 − 0.6 = 4.4 V.

Voltage across three resistors in parallel = 2 A × 1 Ω = 2 V.

Voltage across two resistors in parallel = 2 A × 1.2 Ω = 2.4 V.

$I_1 = 2\,\text{V}/2\,\Omega = 1\,\text{A}$.

$I_2 = 2\,\text{V}/3\,\Omega = 0.67\,\text{A}$.

$I_3 = 2\,\text{V}/6\,\Omega = 0.33\,\text{A}$.

$I_4 = 2.4\,\text{V}/2\,\Omega = 1.2\,\text{A}$.

$I_5 = 2.4\,\text{V}/3\,\Omega = 0.8\,\text{A}$.

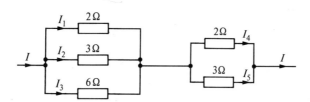

Fig. 2.2.

12 D.C. circuits

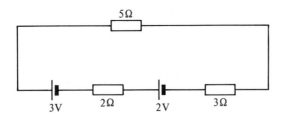

Fig. 2.3.

2.2 The principle of superposition

If two batteries are connected in series the e.m.f.s, and also the internal resistances, are added together. This must be so because, being in series, the same current flows in each. Thus Fig. 2.3 shows two in-series batteries of e.m.f. 3 V and 2 V and internal resistance 2 Ω and 3 Ω respectively connected to a 5 Ω resistor. The total e.m.f. is 5 V and the total resistance in the circuit is 10 Ω leading to a current of 5 V/10 Ω or 0.5 A. This is a fairly easy calculation, but when we consider the case shown in Fig. 2.4, where the batteries are in parallel, such simple ideas can no longer be applied. For example the internal resistances are certainly *not* in parallel because they will *not* have the same voltage across them. There are many ways of solving this problem, one of which is to use the principle of superposition. Another method will be seen later in this chapter.

The principle of superposition states that 'in a *linear* circuit containing more than one source of e.m.f., the current in any component is the algebraic sum of the currents which would flow due to each source of e.m.f. acting alone, all other sources of e.m.f. being replaced by their internal resistances.'

Thus consider the 3 V battery of Fig. 2.4 acting alone, the 2 V battery being replaced by its internal resistance of 3 Ω (Fig. 2.5).

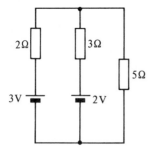

Fig. 2.4.

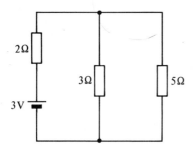

Fig. 2.5.

$3\,\Omega$ in parallel with $5\,\Omega = \dfrac{3 \times 5}{3 + 5} = \dfrac{15}{8} = 1.875\,\Omega.$

Total resistance $= 2 + 1.875 = 3.875\,\Omega.$

Total current from battery $= 3\,\text{V}/3.875\,\Omega = 0.774\,\text{A}.$

Voltage across parallel combination $= 1.875\,\Omega \times 0.774\,\text{A} = 1.45\,\text{V}.$

Current in $5\,\Omega = I_1 = 1.45/5 = 0.29\,\text{A}.$

Now consider the 2 V battery alone (Fig. 2.6). The $2\,\Omega$ and $5\,\Omega$ resistors are in parallel, although it may be more obvious if the circuit is redrawn as in Fig. 2.7.

$2\,\Omega$ in parallel with $5\,\Omega = \dfrac{2 \times 5}{2 + 5} = \dfrac{10}{7} = 1.43\,\Omega.$

Total resistance $= 3 + 1.43 = 4.43\,\Omega.$

Current from battery $= 2\,\text{V}/4.43\,\Omega = 0.45\,\text{A}.$

Voltage across parallel resistors $= 0.45\,\text{A} \times 1.43\,\Omega = 0.64\,\text{V}.$

Current in $5\,\Omega = I_2 = 0.64/5 = 0.13\,\text{A}.$

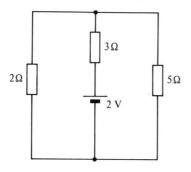

Fig. 2.6.

14 D.C. circuits

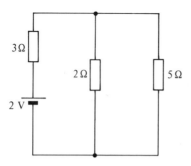

Fig. 2.7.

Using the principle of superposition, the current in the 5 Ω resistor with both batteries present will be

$$I = I_1 + I_2 = 0.29 + 0.13 = 0.42 \text{ A}.$$

Examples 2.2

1. In Fig. 2.4 find the current in each battery.

Solution

Consider the 3 V battery (Fig. 2.5), the current in the battery is 0.774 A upwards. Hence the current in the 3 Ω resistor is

$$0.774 - 0.29 = 0.484 \text{ A downwards}.$$

Now consider the 2 V battery (Fig. 2.7), the current in the battery is 0.45 A upwards. Hence the current in the 2 Ω resistor is

$$0.45 - 0.13 = 0.32 \text{ A downwards}.$$

Hence, using the principle of superposition, the total current (in Fig. 2.4) in the 3 V battery is the algebraic sum of 0.774 A upwards and 0.32 A downwards, or 0.454 A upwards. Similarly the current in the 2 V battery is the sum of 0.45 A upwards and 0.484 A downwards, or 0.034 A downwards. (The 2 V battery is in fact being charged by the 3 V battery.)

2. Find the current in the 10 Ω resistor in the circuit of Fig. 2.8.

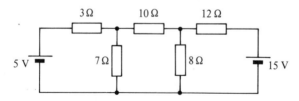

Fig. 2.8.

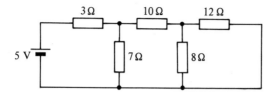

Fig. 2.9.

Solution

Consider firstly the 5 V battery alone (Fig. 2.9).

12 Ω in parallel with 8 Ω = $\dfrac{12 \times 8}{12 + 8}$ = $\dfrac{96}{20}$ = 4.8 Ω.

4.8 Ω in series with 10 Ω = 14.8 Ω.

14.8 Ω in parallel with 7 Ω = $\dfrac{7 \times 14.8}{7 + 14.8}$ = 4.75 Ω.

4.75 Ω in series with 3 Ω = 7.75 Ω.

Current from battery = 5 V/7.75 Ω = 0.65 A.

Voltage across 3 Ω = 3 × 0.65 = 1.95 V.

Voltage across 7 Ω = 5 − 1.95 = 3.05 V.

Current in 7 Ω = 3.05/7 = 0.44 A.

Current in 10 Ω = 0.65 − 0.44 = 0.21 A to the right.

Similarly, consider the 15 V battery alone (Fig. 2.10).

3 Ω in parallel with 7 Ω = $\dfrac{3 \times 7}{3 + 7}$ = 2.1 Ω.

2.1 Ω in series with 10 Ω = 12.1 Ω.

12.1 Ω in parallel with 8 Ω = $\dfrac{12.1 \times 8}{12.1 + 8}$ = 4.82 Ω.

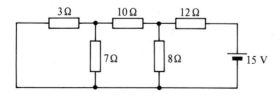

Fig. 2.10.

16 D.C. circuits

4.82 Ω in series with 12 Ω = 16.82 Ω.

Current from battery = 15/16.82 = 0.89 A.

Voltage across 12 Ω = 12 × 0.89 = 10.68 V.

Voltage across 8 Ω = 15 − 10.68 = 4.32 V.

Current in 8 Ω = 4.32/8 = 0.54 A.

Current in 10 Ω = 0.89 − 0.54 = 0.35 A to the left.

Hence, by the principle of superposition, the current flowing in the 10 Ω resistor of Fig. 2.8 is

0.35 − 0.21 = 0.14 A to the left.

2A Test questions

1. Three resistors of values 2 Ω, 4 Ω, and 1 Ω respectively are connected in parallel. Find the total conductance and the total current taken if the parallel combination is connected to a 2 V battery.

2. Find the current taken by each resistor in question 1.

3. Three resistors are connected as shown in Fig. 2.11. Find all the currents and the voltage across the 4 S resistor working entirely in conductances.

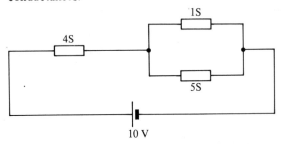

Fig. 2.11.

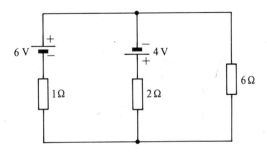

Fig. 2.12

4. State the principle of superposition.
Find the current in the 6 Ω resistor in Fig. 2.12 using the principle of superposition.

2.3 Kirchhoff's Laws

Kirchhoff's laws are two of the most fundamental laws in electrical engineering. They apply both to linear and non-linear circuits, and may be used in a.c. as well as d.c. circuits. There are many ways of stating them—the important thing is to understand what they mean.

Kirchhoff's First Law

The total current flowing towards a junction in a circuit is equal to the total current flowing away from it.

Thus, in Fig. 2.13.

(2.5) $$I_1 + I_2 + I_3 = I_4 + I_5.$$

Sometimes it is stated that the total current flowing *towards* a point in the circuit is zero, currents flowing *away* from the point being regarded as negative. Hence

(2.6) $$I_1 + I_2 + I_3 - I_4 - I_5 = 0$$

which is clearly the same as eqn (2.5).

If this law were not true it would lead to an accumulation of, or a removal of electrons from the point. It follows that if we know all but one of the currents flowing towards or away from a point we can find the one we do not know. In Fig. 2.14

$$I = 3 + 4 + 2 - 8 = 1 \text{ A},$$

and in Fig. 2.15.

$$I = 3 + 4 - 7 = 0.$$

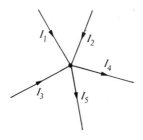

Fig. 2.13

18 D.C. circuits

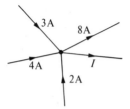

Fig. 2.14

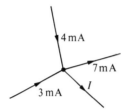

Fig. 2.15

Kirchhoff's Second Law

In any closed loop in a network the algebraic sum of the voltages dropped equals the algebraic sum of the sources of e.m.f. Consider Fig. 2.16 which shows *part* of a network. The voltages dropped in the 2 Ω, 3 Ω, and 4 Ω resistors are 4 V, 3 V, and 2 V respectively, a total of 9 V. The total applied e.m.f. in the closed loop must also be 9 V. Note that Kirchhoff's First Law has been used to determine the currents in the wires leading to other parts of the circuit.

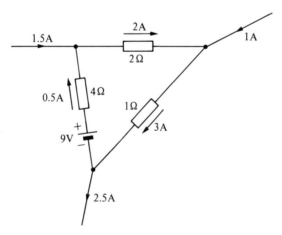

Fig. 2.16.

Kirchhoff's Laws 19

The law simply means that if we start at any point and traverse a complete path we end back at the same potential as that at which we started—obvious because a point cannot be at two different potentials at the same time.

Let us use the laws to solve the circuit of Fig. 2.17 which is Fig. 2.4 redrawn. As we do not know any of the currents we will assume that the currents in the batteries are I_1 and I_2 as shown. Although we could call the current in the 5 Ω resistor I_3 it is marked at $(I_1 + I_2)$, found by the application of Kirchhoff's First Law to point X (or Y).

There are three complete loops in the circuit, ABCDA, AXYDA, and XBCYX. We can solve the circuit by applying Kirchhoff's Second Law to *two* of these loops—two because we have *two* unknowns, I_1 and I_2. I have chosen ABCDA and XBCYX. First take ABCDA; there are two resistors in this loop. The 2 Ω resistor drops a voltage of $2I_1$ volts and the 5 Ω a voltage of $5(I_1 + I_2)$ volts. Hence the total voltage dropped in the loop is $2I_1 + 5(I_1 + I_2)$ and this is equated to the total e.m.f. applied in the loop, namely 3 volts. Hence

(2.7) $$2I_1 + 5(I_1 + I_2) = 3.$$

Collecting the Is together gives

(2.8) $$7I_1 + 5I_2 = 3.$$

Similarly loop XBCYX gives

(2.9) $$3I_2 + 5(I_1 + I_2) = 2$$

or, collecting the Is

(2.10) $$5I_1 + 8I_2 = 2.$$

Eqns (2.8) and (2.10) can now be solved. Multiplying eqn (2.8) by 8 and eqn (2.10) by 5 gives

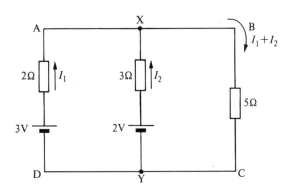

Fig. 2.17.

$$56I_1 + 40I_2 = 24,$$

and $25I_1 + 40I_2 = 10.$

Subtracting

$$31I_1 = 14$$

$$I_1 = 14/31 = 0.452 \text{ A as found previously}.$$

From eqn (2.8)

$$5I_2 = 3 - 7I_1$$
$$= 3 - 3.17 = -0.17$$
$$I_2 = -0.17/5 = -0.034 \text{ A},$$

i.e. it is 0.034 A but in the opposite direction to that marked in Fig. 2.17 in other words *downwards*.

Lastly the current in the 5 Ω resistor is found as

$$I_1 + I_2 = 0.451 - 0.034 = 0.42 \text{ A}.$$

If the loop AXYDA were used, the equation would be slightly more difficult because of the directions of the currents. Going round the loop, the voltages across the 2 Ω and 3 Ω are in opposition. This may be seen in Fig. 2.18 where signs are marked. Hence the total voltage dropped is

$$2I_1 - 3I_2 \text{ (going clockwise)}.$$

Similarly, round this loop the battery e.m.f.s are in opposition, 3 V trying to circulate current in a clockwise direction and 2 V in an anticlockwise direction. Thus the total clockwise e.m.f. is $3 - 2 = 1$ V, and the equation is

$$2I_1 - 3I_2 = 1.$$

Substitution of the values of I_1 and I_2 in this equation confirm that it is correct.

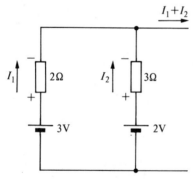

Fig. 2.18.

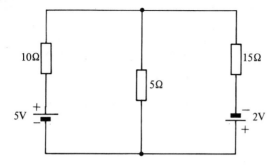

Fig. 2.19.

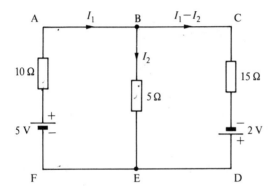

Fig. 2.20.

Examples 2.3

1. Find the current in each resistor of Fig. 2.19.

Solution

The currents are marked in Fig. 2.20.

(2.11)   Loop ABEF $10I_1 + 5I_2 = 5$.

(2.12)   Loop BCDE $15(I_1 - I_2) - 5I_2 = 2$.

Note the sign of the second term.

Eqn (2.12) becomes

(2.13) $15I_1 - 20I_2 = 2$.

Multiplying eqn (2.11) by 4

(2.14) $40I_1 + 20I_2 = 20$.

22 D.C. circuits

Adding eqns (2.13) and (2.14)

$$55I_1 = 22,$$
$$I_1 = 0.4 \text{ A}.$$

From eqn (2.11)

$$5I_2 = 5 - 10I_1,$$
$$= 5 - 4,$$
$$= 1,$$
$$I_2 = 0.2 \text{ A}.$$

The current in the 15 Ω resistor is

$$I_1 - I_2 = 0.4 - 0.2 = 0.2 \text{ A}.$$

2. Repeat question 2 of Examples 2.2 using Kirchhoff's Laws.

Solution

The circuit is redrawn, with currents marked, in Fig. 2.21.

Loop ABGH $3I_1 + 7(I_1 - I_2) = 5,$

$$10I_1 - 7I_2 = 5,$$

(2.15) $$I_1 = 0.5 + 0.7I_2.$$

Loop BCFG $10I_2 + 8(I_2 - I_3) - 7(I_1 - I_2) = 0,$

(2.16) $$-7I_1 + 25I_2 - 8I_3 = 0.$$

Loop CDEF $12I_3 - 8(I_2 - I_3) = -15,$

$$-8I_2 + 20I_3 = -15,$$

(2.17) $$I_3 = -0.75 + 0.41I_2.$$

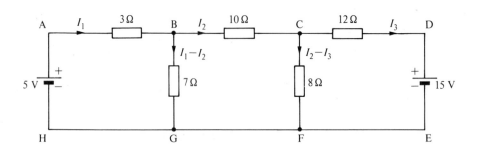

Fig. 2.21.

Substituting eqns (2.15) and (2.17) in eqn (2.16)

$$-7(0.5 + 0.7I_2) + 25I_2 - 8(-0.75 + 0.4I_2) = 0,$$
$$-3.5 - 4.9I_2 + 25I_2 + 6.0 - 3.2I_2 = 0,$$
$$16.9I_2 = -2.5,$$

$I_2 = -0.15$ A (i.e. flowing from C to B).

2.4 The maximum power theorem

A real source of voltage, as we have seen, possess internal resistance. This applies to batteries and rotating generators. It also applies to transducers, that is devices which convert one form of energy to another. In this sense batteries and generators are transducers, but the term is usually thought of in terms of measuring devices. Thus an accelerometer is a device mounted on a moving part which gives an electrical output proportional to the acceleration of the moving part. (Some types of record player pick-up cartridges are, in fact, accelerometers.) Many accelerometers produce only a small amount of electrical power, and we need to transfer as much of that power as possible to the measuring device or amplifier.

Consider a voltage source V with an internal resistance R_g feeding a load R_L (Fig. 2.22). R_L may have any value from zero (short-circuit) to infinity (open-circuit). The power in the load is $I^2 R_L$ and I will have a value between V/R_g, if R_L is zero, to zero if R_L is infinite. In both these extreme cases the power is zero because either I or R_L is zero.

Let us take, as an example, the case of $V = 100$ V and $R_g = 10\,\Omega$. Table 2.1 shows values of I and the power for different values of R_L. It is clear that the power in the load is greatest when the load resistance equals

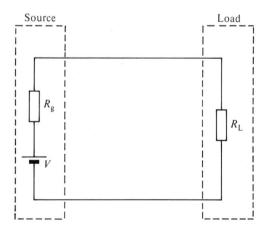

Fig. 2.22.

24 D.C. circuits

TABLE 2.1

$R_L(\Omega)$	TOTAL RESISTANCE $R(\Omega)$	$I = \dfrac{V}{R}$ (A)	$P = I^2 R$ (W)
0	10	10	0
1	11	9.09	82.6
5	15	6.67	222.4
8	18	5.56	247.3
9	19	5.26	249.0
10	20	5.00	250.0
11	21	4.76	249.2
12	22	4.55	248.4
15	25	4.00	240.0
20	30	3.33	221.8
50	60	1.67	139.4
∞	∞	0	0

the resistance of the generator. This is the *maximum power theorem*. It is seen, therefore, that to transfer as much power as possible from a source to a load we must make the load resistance equal to that of the source, or *match* the load to the source. Of course, if the source and load resistances are equal, as much power is developed, or wasted, in the source as is developed in the load. This means that the efficiency of such a system is 50 per cent, not very good. Nevertheless, there is no way in which a given generator can supply a load with more power.

2B Test questions

1. State Kirchhoff's Laws.
Solve question 4 of test questions 2A using Kirchhoff's Laws.

2. Find the currents in the batteries in Fig. 2.12.

3. In Fig. 2.23 V_1 has an e.m.f. of 2 V and an internal resistance of 1 Ω and V_2 has an e.m.f. of 3 V and an internal resistance of 1.5 Ω. Find the current in the 3 Ω resistor using
   (a) the principle of superposition,
   (b) Kirchhoff's Laws.

2.5 Thévenin's Theorem

This is one of the most important, and most used theorems in electrical engineering. Before having a formal statement of the theorem it is worth considering just what Thévenin was really saying.

Consider a 'box' containing a number of linear resistors and sources of e.m.f. (Fig. 2.24). There are two wires coming from the 'box'. Thévenin said that one can put a single source of e.m.f., V_{oc}, and a single resistor, R_{oc}, in series (Fig. 2.25) such that, if the values are correctly chosen, the

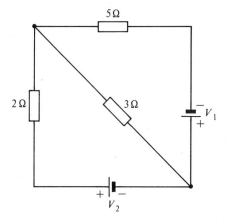

Fig. 2.23.

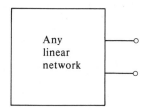

Fig. 2.24.

two 'boxes' are indistinguishable from the outside, i.e. any measurement made on both boxes will give the same result.

Now there are, in fact, very few *different* measurements which can be made. Firstly we can measure the voltage across the terminals on open circuit, V_{oc}. Obviously in Fig. 2.25 (the Thévenin equivalent circuit) this will be the voltage of the e.m.f. source as R_{oc} drops no voltage on open circuit. Secondly we can measure the resistance between the terminals on open circuit, R_{oc}, assuming that all e.m.f. sources have been replaced by their internal resistances. This will be the series resistance in the Thévenin

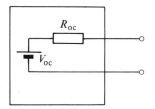

Fig. 2.25.

circuit. A third measurement one could make is the current which would flow in a short circuit connected between the terminals, I_{sc}. However, Fig. 2.25 shows that this current could be obtained from the other two measurements

(2.18) $$I_{sc} = \frac{V_{oc}}{R_{oc}}.$$

Using Thévenin's circuit, it is now possible to calculate the current which would flow if any value of resistance were connected across the terminals. If this resistance is R the current will be given by

(2.19) $$I = \frac{V_{oc}}{R_{oc} + R}$$

as can be clearly seen from Fig. 2.25. Thévenin's Theorem says that this is also the current which will flow in the circuit of Fig. 2.24 if a resistor R is connected across its terminals.

The idea of replacing any linear circuit by the circuit of Fig. 2.25 is an extremely useful concept. In electronic engineering the output of an amplifier working on the linear part of its characteristics is often replaced by the circuit of Fig. 2.25.

The theorem may be stated as follows: The current flowing in a resistor R connected across two points X and Y of an active, linear network is

$$\frac{V_{oc}}{R_{oc} + R},$$

where R_{oc} is the resistance measured across X and Y with R disconnected (all e.m.f. sources being replaced by their internal resistances) and V_{oc} the voltage across X and Y with R disconnected.

Examples 2.4

1. Find the current in the 5 Ω resistor in the circuit of Fig. 2.26. (This is the same circuit as Figs. 2.4 and 2.17.)

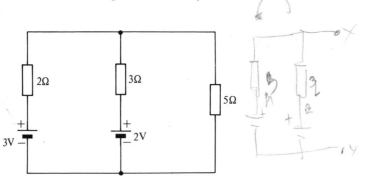

Fig. 2.26.

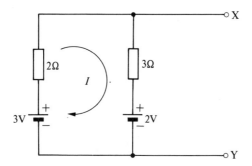

Fig. 2.27.

Solution

Fig. 2.27 shows the circuit redrawn with the 5 Ω resistor removed. A current I will flow, as shown, equal to

$$I = \frac{3-2}{5} = 0.2 \text{ A.}$$

Note
(a) The 3 V and 2 V supplies are opposing each other as far as this current is concerned.
(b) This current of 0.2 A will not flow when the 5 Ω resistor is reconnected.

Now 0.2 A in 2 Ω drops 0.4 V so that the open-circuit voltage across X and Y, V_{oc}, is

$$V_{oc} = 3 - 0.4 = 2.6 \text{ V.}$$

This could also have been found as

$$V_{oc} = 2 + (0.2 \times 3) = 2.6 \text{ V.}$$

The open circuit resistance, R_{oc}, between X and Y with the batteries replaced by short circuits (assuming that their internal resistances are both zero) is 2 Ω in parallel with 3 Ω or

$$R_{oc} = \frac{2 \times 3}{2 + 3} = 1.2 \text{ Ω}$$

so that Fig. 2.27 becomes Fig. 2.28.
If the 5 Ω resistor is now put back, the current in it is given by

$$\frac{2.6}{5 + 1.2} = 0.42 \text{ A.}$$

2. Repeat question 2 of examples 2.2 using Thévenin's Theorem.

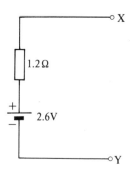

Fig. 2.28

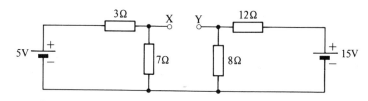

Fig. 2.29.

Solution

Fig. 2.29 shows Fig. 2.8 redrawn with the 10 Ω resistor removed. By the potential divider rule, the potential at X is

$$\frac{7}{10} \times 5 = 3.5 \text{ V}.$$

Similarly the potential at Y is

$$\frac{8}{20} \times 15 = 6.0 \text{ V}.$$

Hence the potential difference between X and Y is 2.5 V (Y positive).

If the batteries are shorted out (Fig. 2.30) the resistance between X and Y is

$$\frac{3 \times 7}{3 + 7} + \frac{8 \times 12}{8 + 12}$$

$$= \frac{21}{10} + \frac{96}{20}$$

$$= 2.1 + 4.8 = 6.9 \, \Omega.$$

The Thévenin equivalent circuit is shown in Fig. 2.31. The current in a 10 Ω resistor connected between X and Y is

$$\frac{2.5}{10 + 6.9} = 0.15 \text{ A (Y to X)}.$$

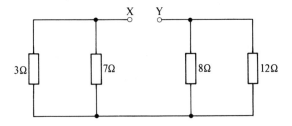

Fig. 2.30

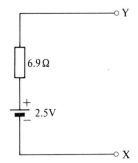

Fig. 2.31.

3. Find the Thévenin equivalent circuit of Fig. 2.32.

Solution

Firstly consider the circuit to the left of AB, Fig. 2.33.

$$I = \frac{5}{5} = 1 \text{ A}.$$

$$V_{AB} = 10 - 3 = 7 \text{ V}.$$

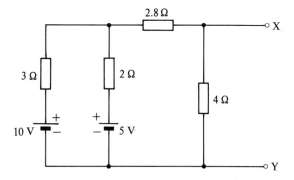

Fig. 2.32.

30   D.C. circuits

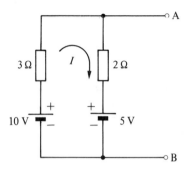

Fig. 2.33.

$$R_{AB} = \frac{2 \times 3}{2 + 3} = 1.2 \, \Omega.$$

Hence the circuit of Fig. 2.32 becomes that of Fig. 2.34. 1.2 Ω in series with 2.8 Ω is 4.0 Ω.

$$R_{XY} = 2 \, \Omega \text{ (4 } \Omega \text{ in parallel with 4 } \Omega\text{)}.$$

$$V_{XY} = \frac{4}{8} \times 7 = 3.5 \text{ V by the potential divider rule.}$$

Hence the Thévenin equivalent circuit, Fig. 2.35.

4. Find the current which would flow in a short circuit across X and Y in Fig. 2.32.

Solution

From the equivalent circuit of Fig. 2.35 this current will be

$$I_{sc} = \frac{3.5}{2} = 1.75 \text{ A}.$$

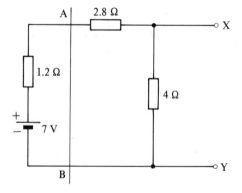

Fig. 2.34.

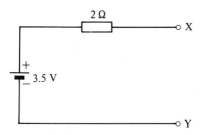

Fig. 2.35

2C Test questions

1. Repeat question 3 of Test questions 2B using Thévenin's Theorem.

2. Find the current in the 5 Ω resistor of Fig. 2.36 using Thévenin's Theorem.

3. Find the value of resistance which, when connected across AB in Fig. 2.37 would receive maximum power. What is this power?

4. Find the current in the 20 Ω resistor of Fig. 2.38 using Thévenin's Theorem.

5. Find, using Thévenin's Theorem, the current in the 1.5 Ω resistor of Fig. 2.39.

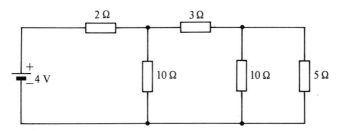

Fig. 2.36

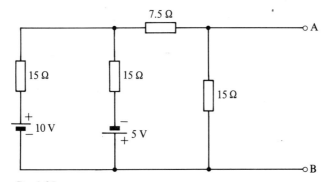

Fig. 2.37.

32 D.C. circuits

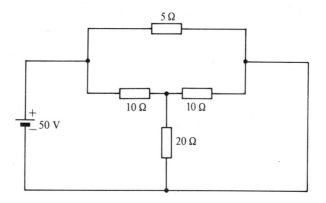

Fig. 2.38.

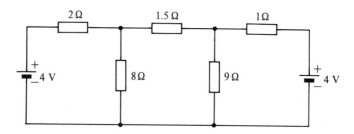

Fig. 2.39.

3 Capacitors and capacitance

3.1 Electric charge and field

The ancient Greeks found that amber rubbed with silk will attract light objects. In fact, the word electricity comes from the Greek word for amber—*electron*. Experiments in the eighteenth century showed that other substances display a similar phenomenom, such as glass rubbed with silk and ebonite rubbed with fur. The substances are said to acquire an electric charge, and it was soon discovered that two kinds of charge existed. Two charged glass rods are found to repel each other as do two charged ebonite rods. However, charged ebonite and glass rods attract one another, leading to the conclusion that

> *like charges repel,*
> *unlike charges attract.*

In about 1750 Benjamin Franklin called the charge possessed by the glass *positive* and that possessed by the ebonite *negative*. He thought of the charging process as a flow of some kind of fluid. We now know that it is, in fact, due to a flow of elementary particles called electrons.

An atom, in very simple terms, may be thought of as a nucleus with a positive charge surrounded by orbiting negatively charged electrons. A body with a negative charge has a surplus of electrons; one with a positive charge a deficit.

We have seen that a charged body has the property of attracting or repelling another charged body situated near to it. The region where this 'invisible' force exists is known as an *electric field* and its effect is rather like that of the *gravitational field* around a mass, such as the earth. The gravitational field has an effect on a mass, the electric field has an effect on a charge. In Chapter 4 we shall discuss the *magnetic field* which has a similar effect on a magnetic pole.

The unit of electric charge is the *coulomb* (C). It is the charge passing a point in one second when a current of 1 ampere is flowing. The ampere is defined as 'that current which when flowing in each of two infinitely long parallel conductors, situated in a vacuum and separated 1 metre between centres, produces on each conductor a force of 2×10^{-7} newton per metre length'.

Electric fields are drawn as lines eminating from a charge and showing the path that would be taken by a free charge moving in the field. By convention, the direction of these lines, shown by arrowheads, is the direction in which a positive charge would move, i.e. the arrows point from the positive charge to the negative charge.

The so-called lines of *electric flux* associated with a positively charged sphere are shown in Fig. 3.1. The field between two unlike charges is

34   *Capacitors and capacitance*

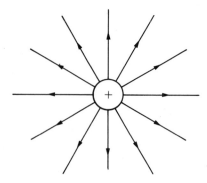

Fig. 3.1. Field round a charged sphere.

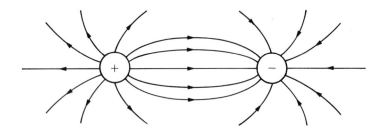

Fig. 3.2.

shown in Fig. 3.2 and between two like charges in Fig. 3.3. One can 'feel' the repulsion in Fig. 3.3 and the attraction in Fig. 3.2.

Potential

When a charge moves in an electric field, work is done. The charge does work if it moves in the direction of the force acting on it; work is done on the charge if it is moved against the force.

The work done in moving a unit positive charge (i.e. a charge of one

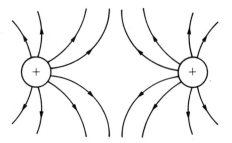

Fig. 3.3.

Electric charge and field

coulomb) from an infinite distance to a point is called the *potential* of the point. If the work done is measured in joules, the potential will be in volts. We are more often concerned with the *potential difference* (p.d.) between two points, which is the work done in moving unit positive charge from one point to the other.

Electric field strength

The strength of an electric field may conveniently be measured in terms of the force which it exerts on a charge. It is measured as the force exerted on a unit positive charge. The symbol for electric field strength is E.

Consider the field between two parallel charged plates, a distance d apart (Fig. 3.4).

If a unit positive charge moves from plate B to plate A the work done on it is V, the p.d. between the plates. However, the work done can be calculated as force times distance, and as the force on the charge equals the field strength E the work done is Ed joules. Hence

(3.1) $$Ed = V$$

(3.2) $$\therefore \quad E = \frac{V}{d}$$

showing that E is measured in volts per metre. As this is the potential fall per unit distance, E may also be called the *potential gradient* of the field.

3.2 Capacitance

It can be shown that the amount of charge given to a conducting body is proportional to the potential to which it is raised, i.e.

(3.3) $$Q \propto V.$$

(3.4) Hence $$Q = CV.$$

where C is a constant depending on the dimensions of the conductor.

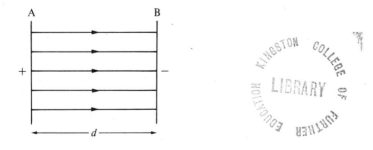

Fig. 3.4.

It represents the charge held by the conductor when it is at unit potential. It is called the *capacitance* of the conductor. If Q is in coulombs and V in volts C will be in *farads* (F). The farad is an extremely large unit, and practical capacitors are measured in microfarads (μF), nanofarads (nF), or picofarads (pF).

The capacitance of isolated conductors is very small, but two conductors placed close together and separated by an insulator called the *dielectric*, have a very much larger capacitance.

A device designed to store charge is called a *capacitor*, the simplest form being two parallel plates as in Fig. 3.4.

A capacitor may be charged by connecting it to a source of e.m.f. such as a battery (Fig. 3.5). Note the symbol for a capacitor, which is drawn like the two parallel plates. When the battery is connected, electrons flow from the top plate of the capacitor anticlockwise round the circuit to the bottom plate. The current in Fig. 3.5 is shown in the conventional direction. This leaves the top plate deficient in electrons (positively charged) and the lower plate with a surplus of electrons (negatively charged).

This flow of electrons continues until the potential difference between the capacitor plates is equal to V, when there is no longer any net e.m.f. round the loop to produce current. The charge on the capacitor must then be CV from eqn (3.4).

If the battery is now removed, the charge remains on the plates and there will still be a p.d. of V volts between them. If a resistor is connected across the capacitor (Fig. 3.6), electrons flow from the negative plate back to the positive one until the p.d. (and hence the charge) is reduced to zero. The process of charging and discharging a capacitor will be dealt with in more detail in Chapter 5.

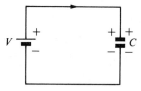

Fig. 3.5.

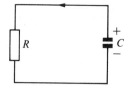

Fig. 3.6.

3.3 Capacitors in parallel and in series

Parallel

Consider two capacitors of capacitance C_1 and C_2 respectively, connected in parallel to a battery of voltage V (Fig. 3.7). The charge on each is given by

$$Q_1 = C_1 V,$$
$$Q_2 = C_2 V.$$

Now imagine that C_1 and C_2 are replaced by a single capacitor of capacitance C which is capable of holding the same total charge, $Q_1 + Q_2$.

Then $CV = Q_1 + Q_2$
$$= C_1 V + C_2 V$$

(3.5) $$\therefore \quad C = C_1 + C_2.$$

Series

If two capacitors in series are connected to a battery V (Fig. 3.8) a current flows round the circuit until the voltage across the two in series is equal to V. As the same current flows in each for the same length of time, the resulting charges must be equal (charge = current × time), say Q.

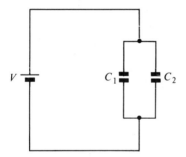

Fig. 3.7. *Capacitors in parallel.*

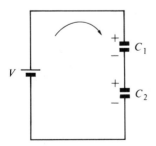

Fig. 3.8. *Capacitors in series.*

Voltage across $C_1 = V_1 = \dfrac{Q}{C_1}$.

Voltage across $C_2 = V_2 = \dfrac{Q}{C_2}$.

If the two in series are equivalent to a single capacitor C, then as

$$V = V_1 + V_2,$$

$$\frac{Q}{C} = \frac{Q}{C_1} + \frac{Q}{C_2}$$

(3.6)   or $\dfrac{1}{C} = \dfrac{1}{C_1} + \dfrac{1}{C_2}.$

3.4 The parallel plate capacitor

Consider the parallel-plate capacitor shown in Fig. 3.9. As two capacitors in parallel have a capacitance equal to the sum of the two (eqn (3.5)) it is clear that the capacitance of the parallel-plate capacitor is directly proportional to the cross-sectional area of the plates, a.

Two equal capacitances in series have a capacitance equal to half of one of them (eqn (3.6)). If we regard the parallel-plate capacitor of Fig. 3.9 as two equal ones in series, each will have plates separated by $d/2$ and we can see that doubling the distance between the plates will halve the capacitance. Hence the capacitance is inversely proportional to the distance between the plates, d.

(3.7) $C \propto \dfrac{a}{d}.$

The actual value of C depends on the material between the plates—the dielectric. On the SI system of units if a is in square metres, d in metres, and C in farads eqn (3.7) becomes

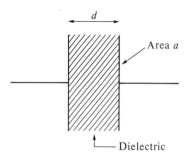

Fig. 3.9. Parallel plate capacitor.

$$C = \frac{8.85 \times 10^{-12} \times a}{d}$$

if the dielectric is a vacuum.

8.85×10^{-12} is a very basic physical quantity in SI units. It is called the *permittivity of free space* and is given the symbol ϵ_0 (ϵ is the Greek letter epsilon). Hence

(3.8) $$C = \frac{\epsilon_0 a}{d}$$

for a vacuum.

If a sheet of insulating material is inserted between the plates the capacitance is found to increase. The capacitance of a capacitor with a certain material as dielectric, relative to that with a vacuum dielectric is known as the *relative permittivity*, ϵ_r, of the material. (It is somtimes called the *dielectric constant* of the material.)

Eqn (3.8) then becomes

(3.9) $$C = \frac{\epsilon_0 \epsilon_r a}{d}.$$

Typical values of ϵ_r are 2-2.5 for paper, 4.5-5.5 for bakelite, and 3-7 for mica. ϵ_r is almost unity for air.

As will be seen in Section 3.7 some capacitors consist of more than two plates (Fig. 3.10). This is effectively seven similar capacitors in parallel, and will have a capacitance seven times that of one of the units. In general if there are n plates there will be $(n-1)$ capacitors and

(3.10) $$C = \frac{\epsilon_0 \epsilon_r a (n-1)}{d}.$$

3.5 Energy stored in a capacitor

Most of us are aware that a charged capacitor stores energy. This can be seen by touching together the leads of a small capacitor that has been charged (a dangerous practice with a large capacitor!).

If we move a charge of Q coulombs from one plate of a capacitor to the other we do work which can be released as energy when it is discharged.

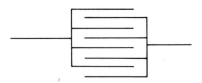

Fig. 3.10.

40 Capacitors and capacitance

At the start of the charging the p.d. between the plates is zero and at the finish it is V volts. The average voltage during charging is thus $V/2$ so that the total work done (and hence the total energy stored) is $V/2$ multiplied by Q or $\frac{1}{2}QV$ joules. As $Q = CV$, this is more frequently written as

(3.11) $\qquad$ energy stored $= \frac{1}{2}CV^2$.

3.6 Dielectric strength

A very important parameter of a capacitor is the voltage which can be applied between the plates without breaking down the dielectric. Clearly, for a given material it is not the voltage itself which matters but the voltage per unit thickness. This is called the *dielectric strength* and is thus the maximum electric field strength or potential gradient which the material can withstand. In practice it will vary with the thickness of the dielectric and with the temperature. Some typical values, in kV/mm, are given below:

Air	3–6
Glass	5–30
Mica	40–200
Paper	4–10
Waxed paper	40–60

Examples 3.1

1. A parallel-plate capacitor has two plates of cross-sectional area 250 cm² separated by 5 mm of air. A p.d. of 1000 V is applied, find
   (a) the electric field strength,
   (b) the capacitance,
   (c) the charge,
   (d) the energy stored.

Solution

(a) $E = \dfrac{V}{d} = \dfrac{1000\,\text{V}}{5\,\text{mm}} = 200\,\text{V/mm or } 200\,000\,\text{V/m}.$

(b) $C = \dfrac{\epsilon_0 a}{d} = \dfrac{8.85 \times 10^{-12} \times 250 \times 10^{-4}}{5 \times 10^{-3}} = 44.25 \times 10^{-12}\,\text{F} = 44.25\,\text{pF}.$

Note: $1\,\text{cm}^2 = 10^{-4}\,\text{m}^2$.

(c) $Q = CV = 44.25 \times 10^{-12} \times 10^3$
$\qquad = 44.25 \times 10^{-9}\,\text{C}.$

(d) Energy $= \frac{1}{2}CV^2 = \frac{1}{2} \times 44.25 \times 10^{-12} \times 10^6$
$\qquad = 22.125 \times 10^{-6}\,\text{J}.$

Dielectric strength 41

2. A capacitor consists of 12 plates each separated by 2 mm of mica of relative permittivity 5. The area of each plate is 200 cm². Find
    (a) the capacitance,
    (b) the voltage across the plates for a charge of 6 × 10⁻⁶ C,
    (c) the energy stored,
    (d) the maximum voltage which can be applied if the dielectric strength of mica is 100 kV/mm.

Solution

(a) $C = \dfrac{\epsilon_0 \epsilon_r (n-1) a}{d}$

$= \dfrac{8.85 \times 10^{-12} \times 5 \times 11 \times 200 \times 10^{-4}}{2 \times 10^{-3}}$

$= 4867.5 \times 10^{-12}$ F $= 4.8675$ nF.

(b) $V = \dfrac{Q}{C}$

$= \dfrac{6 \times 10^{-6}}{4.8675 \times 10^{-9}} = 1233$ V.

(c) Energy $= \tfrac{1}{2}CV^2 = \tfrac{1}{2} \times 4.8675 \times 10^{-9} \times 1233^2$
$= 3.7 \times 10^{-3}$ J.

(d) 100 kV/mm   hence maximum voltage $= 200$ kV.

3. Three capacitors of capacitances 1, 2, and 3 μF respectively are connected
    (a) in parallel,
    (b) in series.
Find the total capacitance in each case.
   If 100 V is applied to each combination find the voltage across and the charge on each capacitor.

Solution

(a) 6 μF total
$Q_1 = C_1 V = 10^{-6} \times 10^2 = 10^{-4}$ C, $V = 100$ V.
$Q_2 = C_2 V = 2 \times 10^{-6} \times 10^2 = 2 \times 10^{-4}$ C, $V = 100$ V.
$Q_3 = C_3 V = 3 \times 10^{-6} \times 10^2 = 3 \times 10^{-4}$ C, $V = 100$ V.

(b) $1/C = 1/C_1 + 1/C_2 + 1/C_3 = 1/1 + 1/2 + 1/3 = 11/6, C = 6/11 = 0.545$ μF.

Q (on each capacitor) $= 0.545 \times 10^{-6} \times 10^2 = 0.545 \times 10^{-4}$ C.

$V_1 = \dfrac{0.545 \times 10^{-4}}{1 \times 10^{-6}} = 54.5$ V.

$$V_2 = \frac{0.545 \times 10^{-4}}{2 \times 10^{-6}} = 27.25 \text{ V}.$$

$$V_3 = \frac{0.545 \times 10^{-4}}{3 \times 10^{-6}} = 18.17 \text{ V}.$$

4. In Fig. 3.11 find
   (a) the total capacitance,
   (b) the charge on, and voltage across each capacitor.

Solution

(a) $2\,\mu\text{F}$ and $4\,\mu\text{F}$ in parallel $= 6\,\mu\text{F}$.

$$6\,\mu\text{F in series with } 3\,\mu\text{F} = \frac{6 \times 3}{9} = 2\,\mu\text{F}.$$

(b) Total charge $= 2 \times 10^{-6} \times 250 = 500 \times 10^{-6}\,\text{C}$.
Charge on $3\,\mu\text{F} = 500 \times 10^{-6}\,\text{C}$.

$$\text{Voltage across } 3\,\mu\text{F} = \frac{500 \times 10^{-6}}{3 \times 10^{-6}} = 166.7 \text{ V}.$$

Voltage across parallel combination $= 250 - 166.7 = 83.3\,\text{V}$.
Charge on $2\,\mu\text{F} = 83.3 \times 2 \times 10^{-6} = 166.7 \times 10^{-6}\,\text{C}$.
Charge on $4\,\mu\text{F} = 83.3 \times 4 \times 10^{-6} = 333.3 \times 10^{-6}\,\text{C}$.

3.7 Practical capacitors

The uses of capacitors are dealt with in other parts of this book and in other areas of your studies of electronics. The type used will depend on the particular application. The main types available are as follows.

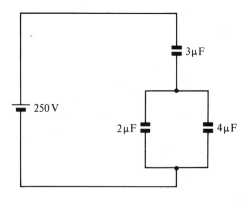

Fig. 3.11.

(i) Paper capacitors

The metal foil (usually aluminium) forming the plates if interleaved with waxed paper and rolled up. They are available from a few nF to tens of μF. Because of their relatively large inductance (see Chapter 4) their use is restricted to low frequencies.

(ii) Mica capacitors

These capacitors are used at high frequencies, such as are encountered in radio and television. They consist of either layers of metal foil separated by layers of mica or are made by coating the two sides of a mica sheet with thin films of silver. They are available up to about 0.01 μF.

(iii) Ceramic capacitors

Silver is deposited on either side of a ceramic disc. They are used for very high frequency work.

(iv) Air capacitors

Used as variable capacitors, they consist of a set of plates which is movable inside a second, or fixed, set. Used for tuning radio sets, etc.

(v) Electrolytic capacitors

These are, in fact, chemical capacitors. Two strips of aluminium foil are interleaved with paper or gauze containing an electrolyte such as ammonium borate. The arrangement is rolled up and packaged in a container. By various processes an oxide insulating film is formed on one plate which acts as the dielectric.

The main advantage of an electrolytic capacitor is that it has a very high capacitance for its physical size. However, electrolytic capacitors have relatively low breakdown voltages and also they can only be used with one plate positive with respect to the other, i.e. they cannot be used where the polarity of one plate might change. One of their most common uses is in d.c. power supplies.

3A Test questions

1. A deficit of electrons produces a positive charge
   (a) true,
   (b) false.

2. Like charges attract
   (a) true,
   (b) false.

3. The work done in moving a unit positive charge from one point in an electric field to another is
   (a) the p.d. between the points,

(b) the difference in charge between the points,
(c) the capacitance between the points,
(d) the work function.

4. Electric field strength, E, is
   (a) the p.d. between two points 1 m apart,
   (b) the force on a unit positive charge,
   (c) the work done in moving unit positive charge from infinity to a point in the field.

5. Field strength, E, is given by
   (a) $E = Vd$,
   (b) $E = V/d$,
   (c) $E = d/V$,
   (d) $E = C/Q$.

6. The relationship between Q, C, and V is
   (a) $Q = CV$,
   (b) $C = QV$,
   (c) $V = QC$.

7. The equivalent of three capacitors in parallel is given by
   (a) $1/C = 1/C_1 + 1/C_2 + 1/C_3$,
   (b) $C = \dfrac{C_1 C_2 C_3}{C_1\ C_2\ C_3}$,
   (c) $C = C_1 + C_2 + C_3$,
   (d) $C = \dfrac{C_1 + C_2 + C_3}{C_1 C_2 C_3}$.

8. The equivalent of three capacitors in series is given by
   (a) $1/C = 1/C_1 + 1/C_2 + 1/C_3$,
   (b) $C = \dfrac{C_1 C_2 C_3}{C_1\ C_2\ C_3}$,
   (c) $C = C_1 + C_2 + C_3$,
   (d) $C = \dfrac{C_1 + C_2 + C_3}{C_1 C_2 C_3}$.

9. Two capacitors of values $10\,\mu F$ and $15\,\mu F$ are connected in series to a d.c. supply. The larger voltage will be across the $10\,\mu F$ capacitor
   (a) true,
   (b) false.

10. The formula for the capacitance of a parallel plate capacitor with n plates, d m apart, and each of cross-sectional area a m^2 is
    (a) $C = \dfrac{\epsilon_0 \epsilon_r (n-1) d}{a}$,
    (b) $C = \dfrac{\epsilon_0 \epsilon_r n d}{a}$,
    (c) $C = \dfrac{a(n-1)}{\epsilon_0 \epsilon_r d}$,
    (d) $C = \dfrac{\epsilon_0 \epsilon_r (n-1) a}{d}$.

where ϵ_0 is the permittivity of free space and ϵ_r is the relative permittivity of the dielectric.

11. The energy stored in a capacitor is
    (a) $\frac{1}{2}CV^2$,
    (b) $\frac{1}{2}QV^2$,
    (c) $\frac{1}{2}VC^2$,
    (d) $\frac{1}{2}\frac{Q}{C}$.

12. A capacitor with a capacitance of 100 µF is connected to a 1500 V d.c. supply. Calculate the charge on the plates.

13. In problem 12 find the energy stored in the capacitor.

14. Three capacitors of values 10 nF, 15 nF, and 25 nF respectively are in parallel. Determine the total capacitance and the total charge if the parallel combination is connected to 750 V d.c.

15. Find the charge on each capacitor in problem 14.

16. Capacitors of values 2 µF, 3 µF, and 6 µF are connected in series to a 200 V d.c. supply. Find
    (a) the total capacitance,
    (b) the charge stored,
    (c) the voltage across each,
    (d) the energy stored in each.

17. A parallel plate capacitor consists of two plates, each 20 cm by 60 cm, separated by glass of thickness 1.5 mm. Find the capacitance if the relative permittivity of glass is 9.

18. A variable (air) capacitor has 20 plates each separated from the next by 2 mm. If the area of overlap of the plates is 250 cm² estimate the capacitance.

19. Find the maximum voltage that can be applied between the plates of the capacitor of problem 17 if the dielectric strength of glass is 20 kV/mm.

20. A capacitor consists of two plates of cross-sectional area 200 cm² separated by 5 mm of air. A p.d. of 200 V is applied between the plates and removed when the capacitor is fully charged. A 5 mm thick block of mica, of relative permittivity 5, is now inserted between the plates. Find the voltage between the plates. (Hint: as there is no external circuit the charge on the plates cannot alter as the mica is inserted.)

4 The magnetic field and electromagnetic induction

4.1 Permanent magnets

Certain iron ores have the property of attracting pieces of iron; one of the first to be discovered was an ore called *magnetite* from which the word *magnet* is derived. A bar of hardened steel rubbed several times with a natural magnet becomes magnetized itself, that is it becomes an artificial magnet. Small pieces of iron are attracted to the ends of a magnet; these areas where the magnetism is concentrated are called *poles*. Every magnet has two different types of pole and it can easily be demonstrated that

like poles repel
unlike poles attract

a law rather like that for electric charges.

If a magnet is freely suspended it will set with one end pointing approximately north, and this end is called the *north pole* the other end being the *south pole*. In fact the earth is a magnetized body with a magnetic south pole situated near the geographical north pole. It is a south pole, of course, because it attracts a magnetic north pole.

The area around a magnet is which a second magnet will experience a force of attraction or repulsion, is called a *magnetic field*. The fields round various combinations of magnets can be plotted using a small compass (a freely suspended magnet). Lines of *force* or *flux* may be drawn showing the direction in which the compass needle sets. By convention arrow heads point from north pole to south pole, that is in the direction in which a free north pole would move if placed in the field. The field round a bar magnet is shown in Fig. 4.1 and that round two bar magnets lying parallel in

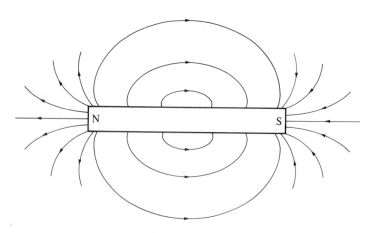

Fig. 4.1. Field round a bar magnet.

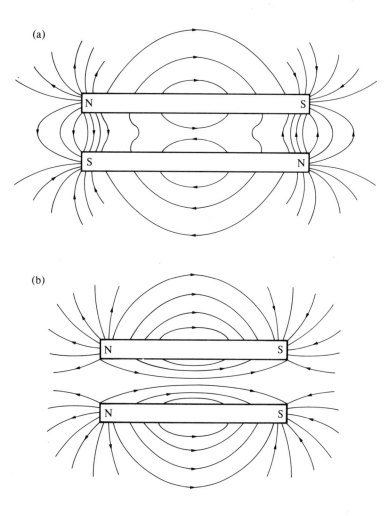

Fig. 4.2.

Fig. 4.2 (a) and Fig. 4.2 (b). The lines of flux also pass through the magnet, although they cannot be plotted there. In fact, a line of flux is a closed loop.

It is not intended in this book to make a detailed study of permanent magnets as it is a rather specialized topic. We shall be more concerned with the magnetic effects of an electric current.

4.2 The magnetic effect of an electric current

It was discovered by Oersted in 1820 that there is a magnetic field round a conductor carrying a current. A compass needle placed underneath a

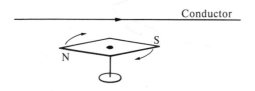

Fig. 4.3.

wire carrying a current, as shown in Fig. 4.3, turns until it is at right angles to the conductor. In the example shown the north pole moves into the paper, the south pole out of it. The force on the needle reverses in direction if the direction of current is reversed or if the compass needle is placed above the conductor.

It can be demonstrated that the field round a straight conductor is in the form of concentric circles with the wire as the centre. Fig. 4.4 shows the field round a conductor with the current going (a) into the paper and (b) out of the paper. Note the conventional signs for current direction.

The direction of the field is given by a rule known as the *right-hand screw rule* or *corkscrew rule*. Imagine that you are driving a right-handed, or normal, screw in the direction of the current. The direction of rotation of the screw is the same as the direction of the magnetic field.

In electrical engineering we are often concerned with coils of wire or *solenoids*. The cross-section of a solenoid is shown in Fig. 4.5, and, as can be seen, the fields round each turn combine to give the effect, outside the solenoid, of a bar magnet. The left-hand end is acting like a south pole and the right-hand end like a north pole. To remember which end produces which polarity imagine that one is viewing the solenoid from that end. From A we see Fig. 4.6 and from B Fig. 4.7. The arrowheads should help your memory!

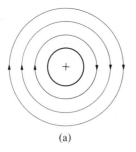
(a)

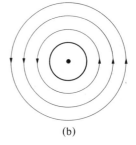
(b)

Fig. 4.4.

Force on a current-carrying conductor

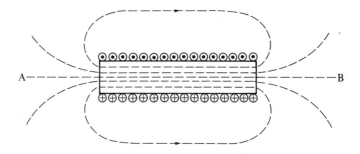

Fig. 4.5.

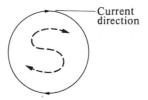

Fig. 4.6.

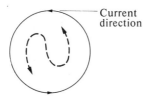

Fig. 4.7.

4.3 Force on a current-carrying conductor

A magnet in a magnetic field experiences a force. As a current-carrying conductor itself possesses a magnetic field it is not surprising to find that such a conductor also experiences a force if situated in a magnetic field. This is, of course, the principle of the electric motor. The force will depend on the angle between the field and the conductor, being greatest when this is a right angle and zero if field and conductor are parallel. If the field and conductor are at right angles the direction of the force will be at right angles to both. The actual direction can be found by using *Fleming's left-hand rule.* The thumb, first finger, and second finger of the left hand are placed mutually at right angles (Fig. 4.8). If the first finger points in the direction of the field, the second finger in the direction of the current then the thumb will point in the direction of the force.

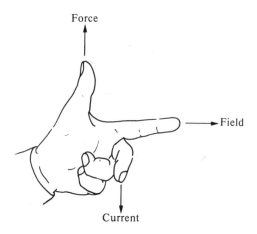

Fig. 4.8. Fleming's left-hand rule.

Alternately, the field due to the current may be superimposed on the other field (Fig. 4.9), aiding above the conductor and opposing below it. Imagining the lines of flux as pieces of stretched elastic shows that the force is downward. (Confirm this by using Fleming's left-hand rule.)

The magnitude of the force can be shown to be proportional to both the current, I, and the length of the conductor, l.

(4.1.) $$F \propto Il.$$

It is also found to depend on the strength of the field. This is called the *flux density* and is given the symbol B. In the SI system of units

(4.2.) $$F = BlI$$

when field and conductor are at right angles. If I is in amperes, l in metres, and B in webers per square metre (Wb/m²) the force F will be in newtons. B is the density of the flux, and the total amount of flux is found by multiplying the flux density by the cross-sectional area. It is, of course, measured in webers (Wb) and is given the symbol ϕ (phi). Thus

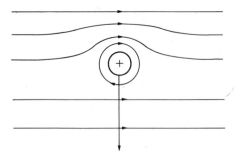

Fig. 4.9.

(4.3) $$\phi = Ba$$

where a is the cross-sectional area. The weber is a very large unit and flux is usually measured in milliweber (mWb) or microweber (μWb).

It should be clear that because a current-carrying conductor has a magnetic field, a force will exist between two current-carrying conductors. If the two conductors are parallel, the force will be of attraction if the currents are flowing in the same direction and repulsion if in opposite directions (Fig. 4.10). This force is used as the definition of the ampere.

The ampere is that current which, when flowing in each of two infinitely long, parallel conductors in a vacuum and separated by 1 m between centres, causes each conductor to have a force exerted on it of 2×10^{-7} newton per metre length of the conductor.

Examples 4.1

1. The flux density inside a solenoid of cross-sectional area 12 cm² is 0.1 Wb/m². Find the total flux in the solenoid.

Solution

$$a = 12 \times 10^{-4} \text{ m}^2$$
$$B = 0.1 \text{ Wb/m}^2$$
$$\therefore \quad \phi = 12 \times 10^{-4} \times 0.1 \text{ Wb}$$
$$= 120 \, \mu\text{Wb}.$$

(a)

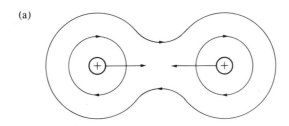

(b)

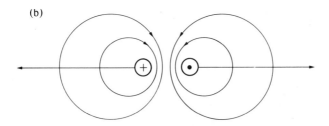

Fig. 4.10.

2. The flux inside a piece of iron is 300 μWb. If the cross-sectional area is 10 cm² find the flux density.

Solution

$$a = 10 \times 10^{-4} \, m^2$$
$$\phi = 300 \times 10^{-6} \, Wb$$
$$B = \frac{300 \times 10^{-6}}{10 \times 10^{-4}} = 0.3 \, Wb/m^2.$$

3. A conductor carrying 800 A is at right angles to a field of flux density 0.15 Wb/m². Find the force on the conductor per metre length.

Solution

$$B = 0.15 \, Wb/m^2$$
$$l = 1 \, m$$
$$I = 800 \, A.$$
$$F = BlI$$
$$= 0.15 \times 1 \times 800$$
$$= 120 \, N.$$

4. The force on a conductor 5 m long in a magnetic field is 750 N. If the conductor is carrying 1200 A find the flux density of the field.

Solution

$$F = 750 \, \dot{N}$$
$$l = 5 \, m$$
$$I = 1200 \, A$$
$$B = \frac{F}{lI} = \frac{750}{5 \times 1200} = 0.125 \, Wb/m^2.$$

4A Test questions

1. Like poles repel
   (a) true,
   (b) false.

2. The earth has a magnetic south pole situated at the geographic north pole
   (a) true,
   (b) false.

3. A compass needle is placed underneath a current carrying conductor as shown in Fig. 4.11. The north pole will be deflected
   (a) into the paper,
   (b) out of the paper,

Fig. 4.11

     (c) to the left,
     (d) to the right.

4. A conductor is carrying a current into the paper (Fig. 4.12). The direction of the magnetic field set up is
     (a) into the paper,
     (b) out of the paper,
     (c) clockwise concentric circles,
     (d) anticlockwise concentric circles.

5. A solenoid carrying a current is shown in Fig. 4.13. If the end shown is a north pole the current will be flowing
     (a) clockwise,
     (b) anticlockwise.

6. If the current carrying conductor shown in Fig. 4.12 is placed in a magnetic field whose lines of force are from right to left, the force on the conductor will be
     (a) to the left,
     (b) to the right,
     (c) upwards,
     (d) downwards.

7. One method of determining the direction of the force in the previous question is to use Fleming's right-hand rule.
     (a) true,
     (b) false.

8. A solenoid of cross-sectional area $10\,\text{cm}^2$ has a total flux passing through it of $90\,\mu\text{Wb}$. Determine the flux density.

9. A conductor 2 m long carrying a current of 500 A is at right angles to a field of flux density $0.2\,\text{Wb/m}^2$. Find the force on the conductor.

Fig. 4.12.

Fig. 4.13.

10. A conductor 2 m long is at right angles to a field of flux density 0.15 Wb/m². If the force on it is 100 N, find the current which it is carrying.

11. Two parallel conductors are carrying currents in the same direction. The force between them is one of
   (a) attraction,
   (b) repulsion.

4.4 Electromagnetic induction

A conductor carrying a current in a magnetic field experiences a force. Conversely, moving a conductor in a magnetic field (by using a force) causes an e.m.f. to be induced, and a current to flow if the circuit is complete. This was discovered by Michael Faraday in 1831. He found that plunging a magnet into a solenoid set up, or *induced*, an e.m.f. in the solenoid. The magnitude of this e.m.f. depends on the strength of the magnet, on the speed of movement, and on the number of turns on the coil. When the magnet is withdrawn, or if the other end of the magnet is used, the e.m.f. has the opposite polarity. An e.m.f. will, of course, also be induced if the magnet is held still and the coil moved. This was one of the most important discoveries in electrical engineering and is the basis of the electric generator.

The same effect is observed if the magnet is replaced by a second solenoid carrying a current. This is not surprising as the magnetic field due to a solenoid is similar to that of a bar magnet. One other fact can be observed. An e.m.f. is induced in the first solenoid not only if it moves relative to the second but also if the current in the second, and hence the strength of its magnetic field, is changed. This is the principle of the transformer and will be dealt with later.

All the causes of the induced e.m.f. discussed above are, in effect, a change in the flux linking the first coil. There is an induced e.m.f. in a circuit whenever the flux linking the circuit is changed. The direction of the induced e.m.f. can be found by a law proposed by Heinrich Lenz. Lenz's Law states that the direction of the induced e.m.f. is such as to oppose the change which is causing it. Hence the magnet plunged into the coil will be repelled. If, for example, the north pole is put in, the e.m.f. will be such that if a current flows it will produce a north pole to repel the magnet (Fig. 4.14). Similarly, when the north pole is removed the current will be in such a direction as to produce a south pole to try to stop the north pole from being removed. The law, clearly, must be true because if the induced current is used to produce energy we must do work in moving the magnet and hence there must be an opposing force.

The e.m.f. induced in the coil is found to be proportional to the number of turns on the coil and on the rate of change of flux linking the coil. Now if the flux linking the coil starts at zero and finishes at a value ϕ, and if this change of flux occurs in t seconds, then

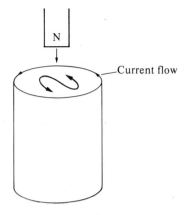

Fig. 4.14.

(4.4) $$E \propto \frac{N\phi}{t}$$

where E is the average induced e.m.f. In fact, if ϕ is in webers and E in volts

(4.5) $$E = \frac{N\phi}{t}.$$

Thus if the flux in a coil of ten turns changes by 0.02 Wb in 1 ms,

$$\text{average e.m.f. } E = \frac{10 \times 0.02}{0.001} = 200 \text{ volts.}$$

Now consider a long, straight conductor moving at right angles to a field of flux density B Wb/m² (Fig. 4.15). If it is moving at v m/s and is of length l m it sweeps out an area of lv m²/s, and hence it cuts Blv webers of flux per second. As can be seen, the conductor and voltmeter form a coil of one turn so that, from eqn (4.5)

(4.6) $$E = Blv \text{ volts.}$$

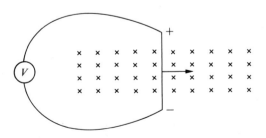

Fig. 4.15.

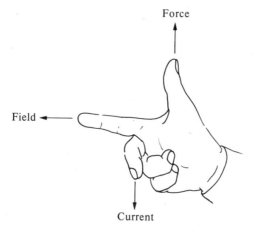

Fig. 4.16. Fleming's right-hand rule.

The direction of the induced e.m.f. can be determined by using Lenz's Law and assuming that the ends of the conductor are connected so that current flows. However, it can also be found by using *Fleming's right-hand rule* (Fig. 4.16). Note that the thumb and fingers stand for the same quantities as in the left-hand rule. If the field in Fig. 4.15 is going into the paper, the e.m.f. will be induced so that the top of the conductor is positive with respect to the bottom and current flows downwards.

Examples 4.2

1. A coil of 50 turns has a flux passing through it of 0.02 Wb. If the direction of the flux is reversed in 0.5 ms find the value of the induced e.m.f.

Solution

The *change* of flux is 0.04 Wb.

Hence
$$E = \frac{N\phi}{t}$$
$$= \frac{50 \times 0.04}{0.5 \times 10^{-3}}$$
$$= 4000 \text{ V}.$$

2. A coil of 100 turns carries a current which produces a flux of 0.01 Wb. The current is halved in 1 ms. Find the induced e.m.f. and state its direction relative to the current.

Solution

$$E = \frac{N\phi}{t} = \frac{100 \times 0.005}{10^{-3}} = 500 \text{ V}.$$

The direction would be in the same sense as the original current, trying to keep it flowing (Lenz's Law).

3. A conductor of length 50 cm is moved at 10 m/s at right angles to a magnetic field. If the induced e.m.f. is 1.5 V find the flux density of the field.

Solution

$$B = \frac{E}{lv} = \frac{1.5}{0.5 \times 10}$$
$$= 0.3 \text{ Wb/m}^2.$$

4. A car travelling at 100 km/h has an axle of length 1.6 m. If the vertical component of the earth's magnetic field is 40 μWb/m^2 find the e.m.f. induced in the axle.

Solution

$$100 \text{ km/h} = \frac{100\,000}{60 \times 60} = 27.8 \text{ m/s}$$

$$E = Blv$$
$$= 40 \times 10^{-6} \times 1.6 \times 27.8$$
$$= 0.00178 \text{ V} = 1.78 \text{ mV}.$$

4B Test questions

1. An e.m.f. of 100 V is induced in a coil when the flux in it falls to zero in 1 ms. Find the e.m.f. induced if the drop in flux occurs in 2 ms.

2. Find the e.m.f. induced in a coil of 1000 turns if the flux in it changes by 10 mWb in a time of 2.5 ms.

3. The flux density in a 200 turn coil of radius 10 cm is 1.2 Wb/m^2. If If this flux falls to zero in 15 ms find the induced e.m.f.

4. A coil of 100 turns has an average e.m.f. of 10 V induced in it when the flux through it is reversed in 1 ms. Find the value of the flux.

5. A conductor of length 100 mm moves at right angles to a magnetic field of flux density 0.2 Wb/m^2. If the e.m.f. induced in it is 2 V find the velocity at which it is moving.

6. A conductor of length 50 cm moves at 50 m/s at right angles to a magnetic field. If the ends of the conductor are connected to a 10 Ω resistor which develops a power of 2.5 W find the flux density of the field.

4.5 Magnetic circuits

Although we are not dealing here with permanent magnets, we are concerned with the lines of flux produced by coils. We have seen that both the force on a current-carrying conductor and the e.m.f. produced when a conductor moves in a magnetic field are proportional to the flux density B, so that motors and generators both depend on the production of large flux densities.

m.m.f.

To set up a current in an electric circuit requires an electromotive force (e.m.f.). By analogy, we speak of a *magnetomotive force* (m.m.f.) being required to set up a magnetic flux. (Note that although we shall continue with this analogy it can be dangerous—magnetic flux is a static thing, unlike a current which is a flow of charge!) This m.m.f. is usually produced by a current in a coil and is proportional to both the current and the number of turns. It is thus convenient to say

(4.7)
$$\text{m.m.f.} = IN$$

where I is the current in amperes and N the number of turns. In SI units it is measured in amperes—because turns has no units. (It used to be called ampere-turns, a much more meaningful name!)

A given m.m.f. will produce very much more flux in a ferromagnetic material, such as iron, than in a non-magnetic material such as air or wood. It is rather like a given e.m.f. producing more current in a conductor than in an insulator.

Consider a coil wound on some material (Fig. 4.17). If the material is iron, the flux produced in it will probably be hundreds of times more than that produced if it were wood. The ratio of m.m.f. to the flux produced is

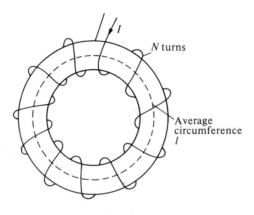

Fig. 4.17.

Magnetizing force and permeability

The *magnetizing force*, *m.m.f. gradient*, or *magnetic field strength*, *H*, which an m.m.f. produces is defined as

(4.8) $$H = \frac{\text{m.m.f.}}{l} = \frac{IN}{l}$$

(see Fig. 4.17). It is measured in amperes per metre and is analogous to electric field strength or potential gradient, *E*, dealt with in the previous chapter.

Now the coil in Fig. 4.17 produces a magnetizing force of IN/l whatever the material used for the former. However, the flux density which it produces is very dependant on the nature of the material. The ratio of the flux density to the magnetizing force, B/H, is called the *permeability* of the material. If the material were a vacuum, the ratio of *B* to *H* is called the *permeability of a vacuum* or the *permeability of free space* and given the symbol μ_0. In SI units its value can be found as follows: Fig. 4.18 shows a long, straight conductor carrying a current of 1 A. A line of flux at a distance of 1 m is shown as a dotted line. As the conductor must have a return path, assumed to be at a great distance, the value of the m.m.f. for the line shown is 1 A and, as the circumference of the circle is 2π m, the magnetizing force is $\frac{1}{2\pi}$ A/m.

Now a second conductor carrying 1 A situated on the circle will experience a force of 2×10^{-7} newtons per metre length, from the definition of the ampere. But from eqn (4.2) this force will be

$$F = BlI$$
$$= B \times 1 \times 1 \text{ for 1 m length.}$$

Hence $B = 2 \times 10^{-7} \text{ Wb/m}^2$.

But $H = \frac{1}{2\pi}$ A/m

so that

(4.9) $$\mu_0 = B/H = \frac{2 \times 10^{-7}}{\frac{1}{2\pi}} = 4\pi\ 10^{-7}. \text{ [SI units]}$$

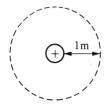

Fig. 4.18.

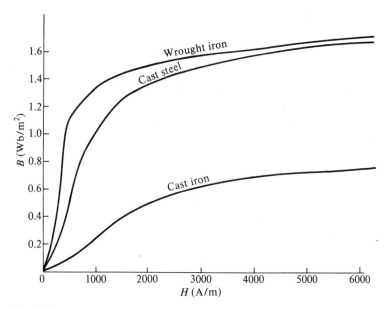

Fig. 4.19.

It is found that the permeability of a non-magnetic substance is so close to μ_0 that this value is used for these substances. However, ferromagnetic materials have permeabilities much higher than μ_0 and it is usual to express the ratio of B to H as

(4.10) $$B/H = \mu_r \mu_0$$

where μ_r is called the *relative permeability*. (A little thought will show that whereas the value of μ_0 depends on the system of units being used, that of μ_r does not!)

The relative permeability of a substance varies with the magnetizing force applied to it, and graphs of B against H for magnetic materials are available. Some examples are shown in Fig. 4.19. It can be seen that after a fairly linear increase of B with H, the curves become almost horizontal. This means that the flux density is hardly increasing even for large increases of magnetizing force. The material is said to be magnetically saturated.

It is frequently useful to have curves showing how the relative permeability, μ_r, varies with either B or H and graphs of these are shown in Figs. 4.20 and 4.21.

Consider, for example, the $B-H$ curve for cast steel (Fig. 4.19).

At $\quad H = 1000$ A/m
$\quad\quad B = 1.0$ Wb/m^2
$\quad\quad B/H = \frac{1}{1000} = \mu_0 \mu_r$

$$\mu_0 = 4\pi\,10^{-7}$$

giving $\mu_r = \dfrac{1}{4\pi\,10^{-4}} = 796$

and this may be found from Fig. 4.21 at $H = 1000\,\text{A/m}$ or Fig. 4.20 at $B = 1.0\,\text{Wb/m}^2$.

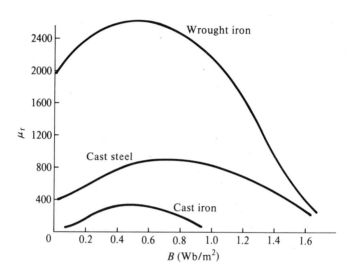

Fig. 4.20.

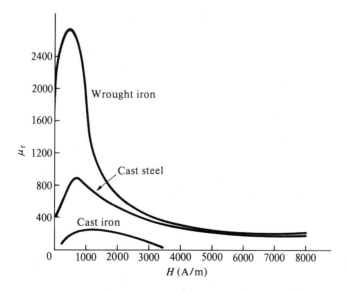

Fig. 4.21.

Reluctance

As has been explained earlier, reluctance, S, is defined as

(4.11)
$$S = \frac{\text{m.m.f.}}{\phi}$$

It is measured in amperes per weber (A/Wb).

However, m.m.f. $= Hl$

and $\phi = Ba$

$\therefore \quad S = \dfrac{Hl}{Ba}.$

But $H/B = 1/\mu_0 \mu_r$

(4.12)
$$S = \frac{l}{\mu_0 \mu_r a}$$

so that the reluctance of a piece of material may be calculated from its dimensions and its permeability.

It can be seen that this formula that eqn (4.11) is not dissimilar to eqn (1.5) giving the resistance of a piece of material in terms of its dimensions and resistivity,

(1.5)
$$R = \frac{\rho l}{a}.$$

Summarizing, a magnetic field is produced by an m.m.f. We also talk of the m.m.f. gradient or magnetizing force, H

(4.7) $\qquad$ m.m.f. $= IN$

(4.8) $\qquad H = $ m.m.f.$/l$.

The total flux produced is ϕ and, perhaps more important, the flux density is B where

(4.3) $\qquad \phi = Ba$.

We can find ϕ or B from H or the m.m.f. if we know the permeability of the material, using either

(4.10) $\qquad B = \mu_0 \mu_r H$

(4.11) or $\qquad \phi = $ m.m.f. $\times S$

(4.12) where $\qquad S = \dfrac{l}{\mu_0 \mu_r a}.$

The equations above enable us to calculate the flux in a material given the m.m.f. or vice versa. Examples 4.3, nos. 1–5 illustrate this. They assume that the material, such as the ring in Fig. 4.17, is homogeneous and does

Magnetic circuits   63

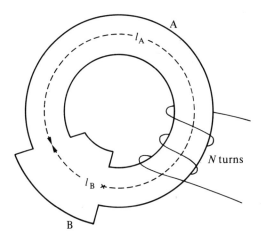

Fig. 4.22.

not change its dimensions. This need not be the case. Fig. 4.22 shows a magnetic circuit comprised of two materials, A and B, of relative permeabilities μ_A and μ_B and cross-sectional areas a_A and a_B respectively. The reluctance of each may be calculated as

$$S_A = \frac{l_A}{\mu_0 \mu_A a_A}$$

and $$S_B = \frac{l_B}{\mu_0 \mu_B a_B}$$

and, like resistances, reluctances in series add, so that the total reluctance S is given by

(4.13) $$S = S_A + S_B.$$

Questions 6 and 7 in Examples 4.3 illustrate problems of this type.

A final, but most important type of problem, is when the ring has a piece of non-magnetic material in it. This might be, for example, an air gap, as in Fig. 4.23. An air gap will be necessary in such devices as motors

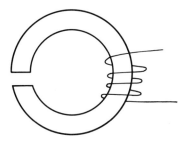

Fig. 4.23.

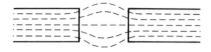

Fig. 4.24.

or generators where a coil has to rotate in a magnetic field. The flux produced by the coil must pass through the air gap because lines of flux are continuous. Whilst the length of the air gap is obvious, its cross-sectional area presents some problems. If the gap is small, it can be assumed that the cross-sectional area is the same as that of the material. However, a larger gap produes fringing or a bulging out of the flux as shown in Fig. 4.24. This reduces the flux density in the gap.

Another effect reducing the flux density in a gap is known as leakage, which occurs because not all the flux produced in the coil is in the material and hence in the gap. Some of it leaks, as shown in Fig. 4.25. The problems in this book assume that the effects of fringing and leakage are negligible.

Fig. 4.25.

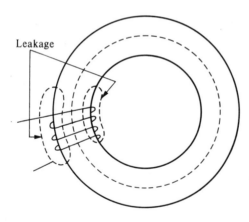

Examples 4.3

1. The coil of Fig. 4.17 consists of 400 turns and carries a current of 4 A. If it is wound on a wooden ring of mean circumference 50 cm and cross-sectional area 6 cm^2 find
    (a) the m.m.f.,
    (b) the magnetizing force,
    (c) the flux density,
    (d) the total flux,
    (e) the reluctance of the ring.

Solution

(a) m.m.f. $= IN$
$= 4 \times 400$
$= 1600$ A.

(b) $H = 1600/0.5 = 3200$ A/m.

(c) $B = \mu_0 H$ ($\mu_r = 1$ for wood)
$= 4 \times \pi \times 10^{-7} \times 3200$
$= 0.004$ Wb/m².

(d) $\phi = B a$
$= 0.004 \times 6 \times 10^{-4}$
$= 2.4 \times 10^{-6}$ Wb $= 2.4 \mu$Wb.

(e) $S = $ m.m.f./$\phi = \dfrac{1600}{2.4 \times 10^{-6}} = 6.6 \times 10^8$ A/Wb

or $S = \dfrac{l}{\mu_0 a}$

$= \dfrac{0.5}{4 \times \pi \times 10^{-7} \times 6 \times 10^{-4}} = 6.6 \times 10^8$ as before.

2. Repeat example 1 if the coil is wound on an iron ring of the same dimensions and assuming that the relative permeability of the iron is 300.

Solution

(a) m.m.f. $= 1600$ A as before.

(b) $H = 3200$ A/m as before.

(c) $B = \mu_0 \mu_r H$
$= 4 \times \pi \times 10^{-7} \times 300 \times 3200$
$= 1.2$ Wb/m².

(d) $\phi = B a$
$= 1.2 \times 6 \times 10^{-4}$
$= 0.00072$ Wb
$= 0.72$ mWb (300 times as much as before).

(e) $S = \dfrac{1600}{0.72 \times 10^{-3}} = 2.2 \times 10^6$ A/Wb.

3. Repeat example 1 for a ring of cast steel, using Fig. 4.19.

Solution

(a) m.m.f. $= 1600$ A as before.

(b) $H = 3200$ A/m as before.

(c) $B = 1.5$ Wb/m² (from Fig. 4.19).

(d) $\phi = 1.5 \times 6 \times 10^{-4}$
    $= 0.0009$ Wb
    $= 0.9$ mWb.

(e) $S = \dfrac{1600}{0.9 \times 10^{-3}} = 1.8 \times 10^6$ A/Wb.

4. A coil of 500 turns is wound on a cast iron former of mean circumference 30 cm and cross-sectional area 5 cm². Find the current required to set up a flux of 300 μWb in the iron.

Solution

$$\phi = 300\,\mu\text{Wb}$$

$$B = \phi/a = \dfrac{300 \times 10^{-6}}{5 \times 10^{-4}} = 0.6\ \text{Wb/m}^2.$$

$$H = 2800\ \text{A/m (from Fig. 4.19)}$$

m.m.f. $= 2800 \times 0.3 = 840$ A

$$I = \tfrac{840}{500} = 1.7\ \text{A}.$$

5. A steel ring has a mean circumference of 20 cm and a cross-sectional area of 10 cm². If μ_r for the steel is 500 find
    (a) the reluctance,
    (b) the flux produced by 2 A in a coil of 100 turns wound on the ring.

Solution

Method 1

(a) $S = \dfrac{l}{\mu_0 \mu_r a}$

$= \dfrac{0.2}{4 \times \pi \times 10^{-7} \times 500 \times 10 \times 10^{-4}}$

$= 0.318 \times 10^6$ A/Wb.

(b) m.m.f. $= 200$ A

$$\phi = \dfrac{200}{0.318 \times 10^6}\ \text{Wb} = 628\,\mu\text{Wb}.$$

Method 2

m.m.f. $= 200$ A

$H = 200/0.2 = 1000$ A/m

$B = 4 \times \pi \times 10^{-7} \times 500 \times 1000 = 0.628\ \text{Wb/m}^2$

Magnetic circuits 67

$$\phi = 0.628 \times 10 \times 10^{-4} \text{Wb} = 628\,\mu\text{Wb}$$

$$S = \frac{200}{628 \times 10^{-6}} = 0.318 \times 10^6 \text{A/Wb}.$$

6. In Fig. 4.22 the ring has the following dimensions: $l_A = 20\,\text{cm}$, $l_B = 10\,\text{cm}$, $a_A = 3\,\text{cm}^2$, and $a_B = 6\,\text{cm}^2$. If $\mu_A = 500$ and $\mu_B = 300$ find the total reluctance and the flux and flux density produced by 4 A flowing in a coil of 200 turns.

Solution

$$S_A = \frac{0.2}{4 \times \pi \times 10^{-7} \times 500 \times 3 \times 10^{-4}} = 1.06 \times 10^6 \text{A/Wb}.$$

$$S_B = \frac{0.1}{4 \times \pi \times 10^{-7} \times 300 \times 6 \times 10^{-4}} = 0.44 \times 10^6 \text{A/Wb}.$$

$$S = (1.06 + 0.44) \times 10^6$$
$$= 1.5 \times 10^6 \text{A/Wb}.$$

m.m.f. $= 4 \times 200 = 800\,\text{A}$.

$$\phi = \frac{\text{m.m.f.}}{S} = \frac{800}{1.5 \times 10^6}\,\text{Wb} = 533\,\mu\text{Wb}.$$

$$B_A = \frac{533 \times 10^{-6}}{3 \times 10^{-4}} = 1.78\,\text{Wb/m}^2.$$

$$B_B = \frac{533 \times 10^{-6}}{6 \times 10^{-4}} = 0.89\,\text{Wb/m}^2.$$

7. In Fig. 4.22 $l_A = 30\,\text{cm}$, $l_B = 5\,\text{cm}$, $\mu_A = 400$, $\mu_B = 250$, $a_A = 4\,\text{cm}^2$, and $a_B = 5\,\text{cm}^2$. If the coil consists of 300 turns, find the current required to produce a flux density of $0.5\,\text{Wb/m}^2$ in material B.

Solution

Method 1

$$S_A = \frac{0.3}{4 \times \pi \times 10^{-7} \times 400 \times 4 \times 10^{-4}}$$
$$= 1.49 \times 10^6 \text{A/Wb}.$$

$$S_B = \frac{0.05}{4 \times \pi \times 10^{-7} \times 250 \times 5 \times 10^{-4}}$$
$$= 0.32 \times 10^6 \text{A/Wb}.$$

$$S = 1.81 \times 10^6 \text{A/Wb}.$$

$B_B = 0.5\,\text{Wb/m}^2$
$\phi = 0.5 \times 5 \times 10^{-4}\,\text{Wb} = 0.25\,\text{mWb}.$

$$S = \text{m.m.f.}/\phi$$

$$\begin{aligned}
\text{m.m.f.} &= \phi S \\
&= 0.25 \times 10^{-3} \times 1.81 \times 10^6 \\
&= 452.5 \text{ A.}
\end{aligned}$$

$$I = 452.5/300 = 1.51 \text{ A.}$$

Method 2

material B

$$B_B = 0.5 \text{ Wb/m}^2.$$

$$H_B = \frac{0.5}{4 \times \pi \times 10^{-7} \times 250} = 1591.5 \text{ A/m.}$$

$$\text{m.m.f.} = H_B l_B = 1591.5 \times 0.05 = 79.58 \text{ A.}$$

material A

$$\phi = 0.5 \times 5 \times 10^{-4} \text{ Wb} = 0.25 \text{ mWb.}$$

$$B_A = \frac{0.25 \times 10^{-3}}{4 \times 10^{-4}} = 0.625 \text{ Wb/m}^2.$$

$$H_A = \frac{0.625}{4 \times \pi \times 10^{-7} \times 400} = 1243.4 \text{ A/m.}$$

$$\text{m.m.f.} = H_A l_A = 1243.4 \times 0.3 = 373.02 \text{ A.}$$

$$\text{Total m.m.f.} = 452.6 \text{ A.}$$

$$I = 452.6/300 = 1.51 \text{ A.}$$

8. A steel ring of cross-sectional area 5 cm² and mean circumference 40 cm is wound with a coil of 200 turns. Find the current required to produce a flux of 800 µWb. Find the current required to produce the same flux if an airgap of length 1 mm is cut in the ring, neglecting leakage and fringing effects.

Solution

$$\phi = 800 \, \mu\text{Wb.}$$

$$B = \frac{800 \times 10^{-6}}{5 \times 10^{-4}} = 1.6 \text{ Wb/m}^2.$$

$$H = 3500 \text{ A/m (from Fig. 4.15)}$$

$$\text{m.m.f.} = 3500 \times 0.4 = 1400 \text{ A.}$$

$$I = \tfrac{1400}{200} = 7 \text{ A.}$$

Magnetic circuits 69

With gap

The m.m.f. required to produce the flux in the steel is still 1400 A (in fact, the length of the steel is reduced by 1 mm but this is insignificant).

for the air gap

$$\phi = 800\,\mu\text{Wb}.$$

$$B = 1.6\,\text{Wb/m}^2.$$

$$H = \frac{1.6}{4 \times \pi \times 10^{-7}} = 1.273 \times 10^6\,\text{A/m}.$$

m.m.f. $= 1.273 \times 10^6 \times 10^{-3} = 1273$ A.

Total m.m.f. $= 2673$ A.

$$I = 2673/200 = 13.37\,\text{A}.$$

4C Test questions

1. A current of 5 A flows in a coil consisting of 200 turns. The m.m.f. produced is
   (a) 1000 A,
   (b) 40 A,
   (c) 0.025 A,
   (d) dependant on the material on which the coil is wound.

2. The flux, ϕ, m.m.f., and reluctance, S are connected by the equation
   (a) m.m.f. $= \phi S$,
   (b) $\phi = $ m.m.f. $\times S$,
   (c) $S = $ m.m.f. $\times \phi$.

3. Magnetizing force, H, m.m.f., and the length of the magnetic circuit are connected by the equation
   (a) m.m.f. $= H \times l$,
   (b) $H = $ m.m.f. $\times l$,
   (c) $l = $ m.m.f. $\times H$.

4. An iron ring has a mean circumference of 1.5 m and a cross-sectional area of 10 cm^2. A coil of 500 turns is wound on the ring and a current of 2 A produces a flux of 1.1 mWb. Find the relative permeability of the iron.

5. A coil of 600 turns is wound on a circular ring (Fig. 4.17) of mean circumference 75 cm and cross-sectional area 12 cm^2. If the relative permeability of the material forming the ring is 250 find the flux produced by a current in the coil of 5 A.

6. Repeat question 5 if the ring were replaced by one made of cast iron (Fig. 4.19).

7. The dimensions of the ring shown in Fig. 4.22 are $l_A = 10$ cm, $l_B = 50$ cm, $a_A = 10$ cm^2, $a_B = 8$ cm^2, $\mu_{rA} = 300$, and $\mu_{rB} = 500$. Find the current needed in a coil of 500 turns to produce a flux density of 0.75 Wb/m^2 in material A.

8. If an airgap of length 2 mm is cut in material B of the ring in the previous question, find the current needed to produce the same flux density in material A.

4.6 Hysteresis

Fig. 4.19 shows the graphs of B against H for various magnetic materials. They are determined by applying *increasing* values of current in a coil would on the material. An example is shown in the curve oa of Fig. 4.26. However, if, when at point a, the current in the coil, and hence the value of H, is *reduced* the value of B changes as shown by the curve ab and not along the original curve. It can be seen that even when the magnetizing force is reduced to zero, there is still a flux density, known as the *remanent flux density*, in the material. The material has become magnetized. If the material had been originally saturated, at point a, the value of the remanent flux density is called the *remanence* of the material. Its value in Fig. 4.26 is given by ob.

To reduce the flux density to zero actually requires a *reversal* of the magnetizing force, as can be seen by the portion of the curve bc. The value of H, in the negative sense, required to reduce B to zero is called the *coercive force* and the particular value when the material had been originally saturated is the *coercivity*. It is oc in Fig. 4.26. A further increase in H in the negative sense will eventually saturate the material in the opposite direction, curve cd. A reduction and reversal of H produces the curve defa.

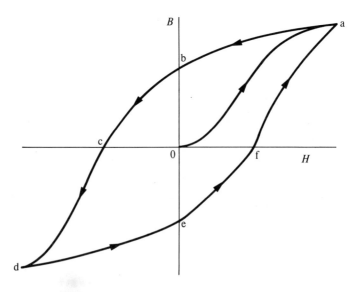

Fig. 4.26.

It is clear that the flux density changes lag behind the magnetizing force changes and the phenomenon is called *hysteresis*, the complete curve of Fig. 4.26 being known as a *hysteresis loop*.

It can be shown that if a magnetic material is taken round a hysteresis loop energy is expended, the material increasing in temperature. The energy used in taking the substance round the loop is, in fact, proportional to the area of the loop.

The subject of hysteresis will be mentioned again in Chapter 8 when dealing with transformers. Transformers consist of coils wound on a magnetic material. When an alternating current is passed through one of the coils, the magnetic material is *taken round* the hysteresis loop once during each cycle of the current. This produces an energy loss, known as the transformer's iron loss, which is clearly proportional to both the area of the loop and the frequency of the supply. The energy loss becomes apparent by the rise in temperature of the magnetic material.

4.7 Self inductance

A current in a circuit such as a coil produces lines of flux. We have seen (Section 4.4) that if this current, and hence flux, is changed, an e.m.f. will be induced in a second coil cut by this flux. This effect is known as *mutual induction*. However, the changing lines of flux are also cutting the original coil and hence will induce an e.m.f. in it. This, by Lenz's Law, will be such as to oppose the change, that is to keep the current steady. This effect is known as *self inductance* and a component possessing self inductance is called an *inductor*. All circuits possess self inductance, although it is often negligible. The circuit symbol for an inductor is shown in Fig. 4.27.

The magnitude of the e.m.f. induced in an inductor is proportional to the rate of change of current. The American Joseph Henry proposed the unit of inductance as the *henry* and it is such that 'a circuit has a self inductance of one henry if a rate of change of current of one ampere per second induces in it an e.m.f. of one volt'. The symbol used for inductance is L and

(4.14)   induced e.m.f. $= L \times$ rate of change of current.

Some books show a minus sign in eqn (4.14) to indicate that the e.m.f. induced in opposing the change of current. The inclusion or omission of the minus sign both have merits. The important fact to remember is that the induced e.m.f. *opposes* the change which is causing it!

Fig. 4.27.

The importance of self inductance can be seen by imagining that, for example, we are breaking a current of 10 A in a coil of 1 H in a time of 1 μs

$$\text{induced e.m.f.} = 1 \times \frac{10}{10^{-6}}$$

$$= 10^7 \text{V}.$$

a value which would clearly cause severe arcing at the switch contacts! Other effects of self inductance in a d.c. circuit will be dealt with in the next chapter.

If a current of I amperes is reduced to zero in t seconds in a coil of inductance L henrys the average e.m.f. induced is, from eqn (4.14)

$$E = L\frac{I}{t}.$$

But eqn (4.5) states that

$$E = \frac{N\phi}{t}$$

and hence

$$\frac{N\phi}{t} = \frac{LI}{t}$$

(4.15)
$$L = \frac{N\phi}{I}.$$

Imagine a coil of N turns wound on a non-magnetic former of length l m and cross-sectional area a m², If it carries a current of I amperes

$$\text{m.m.f.} = IN$$

$$H = \frac{IN}{l}$$

$$B = \frac{\mu_0 IN}{l}$$

$$\phi = \frac{\mu_0 INa}{l}.$$

Substituting this value of ϕ in eqn (4.15)

(4.16)
$$L = \frac{\mu_0 N^2 a}{l}$$

If the coil is wound on a magnetic former and the current is such that the relative permeability is μ_r

(4.17)
$$L = \frac{\mu_0 \mu_r N^2 a}{l}.$$

It is of particular importance to note that

(4.18) $\qquad L \propto N^2.$

The fact, referred to above, that breaking an inductive circuit can produce arcing, shows that the current carrying inductor is storing energy, just as a charged capacitor does. In fact the energy stored in an inductor of inductance L henrys carrying a current of I amperes is given by

(4.19) $\qquad$ energy stored $= \tfrac{1}{2} L I^2$ joules

(compare with $\tfrac{1}{2} C V^2$ of eqn (3.11)).

4.8 Mutual inductance

The henry is also the unit of mutual inductance, the definition now being that 'the mutual inductance between two circuits is one henry if a change of current of one ampere per second in one induces an e.m.f. of one volt in the other'. The symbol for mutual inductance is M, and the circuit symbol is shown in Fig. 4.28.

Consider two coils A and B consisting of N_1 and N_2 turns respectively wound on a magnetic former such that the flux produced by A passes through B (Fig. 4.29). If the current in A is increased so that a flux is set up in t s the induced e.m.f. is

$$e_1 = \frac{N_1 \phi}{t}$$

Fig. 4.28.

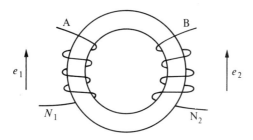

Fig. 4.29.

and this is clearly the e.m.f. which must be applied to cause this change in current. The e.m.f. induced in coil B is

$$e_2 = \frac{N_2 \phi}{t}$$

and hence

(4.20)
$$\frac{e_1}{e_2} = \frac{N_1}{N_2},$$

i.e. the ratio of the induced e.m.f.s is the same as the turns ratio.

Of course the e.m.f.s are only induced whilst the flux, in other words the current, is changing. However, if the current is continually changing, as in an a.c. circuit, alternating e.m.f.s will be induced. This is the principle of the transformer, where an applied e.m.f. e_1 will induce an e.m.f. e_2 which is smaller or greater than e_1 by the turns ratio. Transformers are dealt with in Chapter 8.

Examples 4.4

1. The current in a coil of 5 mH is changed from 100 mA to 250 mA in 30 μs. Find the average e.m.f. induced.

Solution

Average e.m.f. = $L \times$ rate of change of current

$$= 5 \times 10^{-3} \times \frac{(250 - 100) \, 10^{-3}}{30 \times 10^{-6}}$$

$$= 25 \text{ V}.$$

2. A current of 1 A in a coil of 500 turns produces a flux of 2 mWb. Find the self inductance of the coil.

Solution

$$L = \frac{500 \times 2 \times 10^{-3}}{1} = 1 \text{ H}.$$

3. A coil is wound on a non-magnetic former 1 m long and of cross-sectional area 0.8 cm². Find the number of turns required to produce an inductance of 0.1 mH.

Solution

$$N^2 = \frac{lL}{\mu_0 a}$$

$$= \frac{1 \times 0.1 \times 10^{-3}}{4 \times \pi \times 10^{-7} \times 0.8 \times 10^{-4}}$$

$$= 994\,718$$

$$N = 997.$$

4. A reduction of current in a coil A from 0.1 A to zero in 2 ms induces an e.m.f. of 50 V in a coil B. Calculate the mutual inductance between the coils.

Solution

$$\text{e.m.f.} = M \times \text{rate of change of current}$$

$$M = 50 \div \frac{0 \cdot 1}{2 \times 10^{-3}}$$

$$= 1 \text{ H}.$$

5. A coil of inductance 1 H carries a current of 5 A. Find the energy stored.

Solution

$$\text{energy} = \tfrac{1}{2} L I^2$$
$$= \tfrac{1}{2} \times 1 \times 25$$
$$= 12.5 \text{ J}.$$

4D Test questions

1. A coil of 125 turns is wound on a magnetic former of relative permeability 450. If the former is of length 10 cm and cross-sectional area 0.5 cm^2 find the inductance of the coil.

2. How many turns are required if the inductance of the coil in the previous question is to be doubled.

3. The current in the coil of question 1 is changed from 25 mA to zero in 5 μs. Find the average e.m.f. induced.

4. Find the flux produced in the coil of question 1 when it is carrying a current of 25 mA.

5. What is the energy stored by the coil of question 4?

5 Inductance and capacitance in d.c. circuits

Chapter 2 dealt with circuits comprising resistors and batteries and how to calculate the currents flowing in d.c. circuits. These currents, in fact, flow *immediately* the circuit is completed, i.e. if, in Fig. 5.1, the switch S is closed at a time t the current V/R will flow from that time (Fig. 5.2).

In practice, all circuits contain some inductance and some capacitance, although often the effect of one or both of these is negligible. Circuits containing inductance and resistance (where the capacitance is negligible) are often known as *L-R* circuits, those containing capacitance and resistance (inductance being negligible) are called *C-R* circuits. *L-C-R* circuits, which contain all three, will not be dealt with, as far as d.c. is concerned, in this book.

It will be seen later that the effects of inductance and capacitance in a d.c. circuit are transitory, i.e. the effects die out after a certain time. Such effects are known as transients and although only lasting perhaps for a very short time, can be very important in certain instances.

5.1 *L-R* circuits

Even straight pieces of wire posses inductance, though the value may be very small indeed. We are usually concerned with larger values, such as

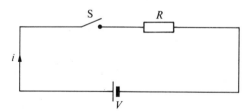

Fig. 5.1.

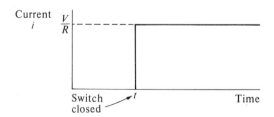

Fig. 5.2. Response of the circuit of Fig. 5.1.

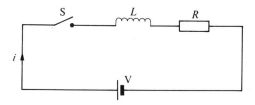

Fig. 5.3.

occur in the windings of machines or the coils used in various electronic circuits. Consider a coil with an inductance L and resistance R connected, via a switch S, to a d.c. supply V (Fig. 5.3). It is convenient to show L and R as separate components, even though in this case R represents the resistance of the coil. The only component offering opposition to the flow of *steady current* is the resistor, and it is reasonable to expect that the current will, at some time, be V/R. Clearly it must be zero before S is closed, so that there must be some rate of change of current as it changes from zero to V/R. Now an inductor has an e.m.f. induced in it when the current changes and this e.m.f. will oppose the change (Lenz's Law). A little thought will show that as the current increases, the voltage across R increases in proportion, so that there is less voltage appearing across L (the sum of the two must equal V (by Kirchhoff's Law)). This can only occur if the rate of change of current is *decreasing*; the graph of current against time will be similar in shape to that of Fig. 5.4. In fact, the current *never* actually reaches the value V/R, although it does to all practical intents and purposes. This type of graph, common in all branches of science and engineering, is called an *exponentially increasing curve*.

Consider the graph at a time t_1 (Fig. 5.5) when the current is i_1. The rate of increase of current at this time is given by the slope of the graph at this point, i.e. the slope of the tangent to the curve, AB:

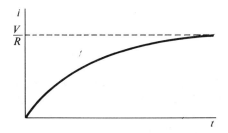

Fig. 5.4. *Response of the circuit of Fig. 5.3.*

78 Inductance and capacitance in d.c. circuits

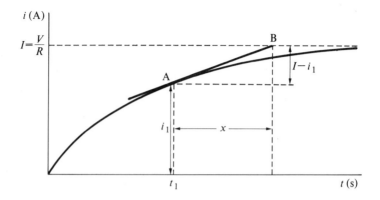

Fig. 5.5.

$$\text{slope of curve at A} = \frac{I-i_1}{x} \text{ A/s,}$$

$$\text{e.m.f. induced in } L \text{ (at A)} = \frac{L(I-i_1)}{x} \text{ V.}$$

Now the rest of the applied voltage, V, must appear across R and be equal to Ri_1:

$$Ri_1 + \frac{L(I-i_1)}{x} = V = IR$$

$$R(I-i_1) = \frac{L(I-i_1)}{x}$$

or

$$x = \frac{L}{R} \text{ s.}$$

This means that the rate of increase of current *at any instant* is such that if that rate were to continue (line AB) the final current (V/R) would be reached in a time L/R seconds. This, of course, applies equally at the instant of switching on, and is shown in Fig. 5.6. L/R is an important quantity called the *time constant* of the circuit, T. It can be seen that if $L = 0$ the time constant would be zero, i.e. the current would reach V/R instantaneously.

The time constant may be defined as the time taken for the current to reach its final value if the original rate of increase were to continue. Of course, the original rate of increase is *not* maintained, and it can be shown that the actual value of the current after one time constant (L/R seconds) is $0.632\,V/R$ (often taken as being about two-thirds of the final value). This leads to another definition of the time constant as the time taken for the current to reach 0.632 of its final value.

After two time constants it will have reached $0.632\,V/R$ plus 0.632 of the difference between its value after one time constant and V/R,

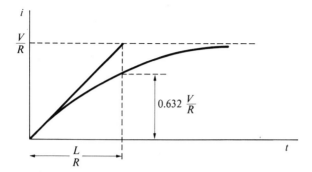

Fig. 5.6.

i.e. $0.632 \frac{V}{R} + 0.632(1 - 0.632)\frac{V}{R}$

$= 0.865 \frac{V}{R}.$

The fraction of V/R reached after various time constants is shown in Table 5.1.

TABLE 5.1

Time in time constants	Fraction of V/R reached
1	0.632
2	0.865
3	0.950
4	0.982
5	0.993
10	0.999 95

After five time constants, for example, it is equal to V/R within 0.7 per cent, good enough for most practical situations!

As an example consider a coil of inductance 2 H and resistance 1 Ω connected to a supply of 10 V. The final current will be 10 A. The time constant is L/R or 2/1 seconds, so that after 2 seconds the current is 6.32 A, after 4 seconds 8.65 A, etc. There is no way in which the increase of current in this coil can be speeded up unless, of course, it is connected to a supply greater than 10 V.

Exponential graphs can be drawn by construction. Consider the example above of a coil of time constant 2 s connected to a 10 V supply (Fig. 5.7 shows the method of construction). The final value of 10 A has been marked by a dotted line. AB is marked onto this and is of a length representing two seconds (the time constant). The point B is joined to the origin 0, and this line is the slope of the graph at the origin. A point C is now taken near to 0 on the line OB. It is joined to D where BD equals horizontal distance between 0 and C. This procedure is repeated and the

80 Inductance and capacitance in d.c. circuits

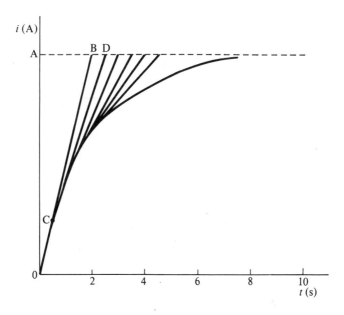

Fig. 5.7.

required curve sketched as shown. The more points taken, of course, the more accurate the curve.

The mathematical derivation of the equation of the curve is somewhat complicated, but the result is that the current i at a time t is given by

(5.1) $$i = \frac{V}{R}\left[1 - \exp\left(-\frac{Rt}{L}\right)\right].$$

If we insert the values of the previous example eqn (5.1) becomes

$$i = 10\left[1 - \exp\left(-\frac{t}{2}\right)\right],$$

hence the current after one second is

$$\begin{aligned}
i &= 10\,[1 - \exp(-\tfrac{1}{2})] \\
&= 10\,(1 - 0.607) \\
&= 10 \times 0.393 \\
&= 3.93 \text{ A},
\end{aligned}$$

and after, say, ten seconds

$$\begin{aligned}
i &= 10\,[1 - \exp(-\tfrac{10}{2})] \\
&= 10\,(1 - 0.000\,05) \\
&= 10 \times 0.999\,95 \\
&= 9.9995 \text{ A}.
\end{aligned}$$

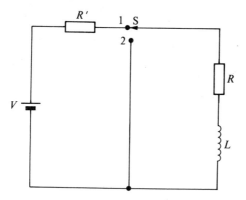

Fig. 5.8.

Decay of current in an L-R *circuit*

We saw in the previous chapter that breaking the current flowing in an inductive circuit can produce very large e.m.f.s which can damage switch contacts. Circuits to large inductive loads are often broken by switching a resistor across the load when the supply is removed, allowing the current to die away more slowly. Fig. 5.8 shows a coil, L and R, with a current V/R flowing (i.e. the switch has been in position 1 for a long time). S is now switched to position 2 (Fig. 5.9). It is assumed that the switch is a make-before-break type, R' being included to protect the battery when it is momentarily shorted out. As the current i starts to fall, an e.m.f. is induced in L trying to maintain its level. This e.m.f. is proportional to the rate of change of current and a graph of current against time will be a falling exponential curve (Fig. 5.10). It can be constructed in a manner similar to Fig. 5.7. If the original rate of decrease were to continue (line AB) the current would fall to zero in a time equal to the time constant,

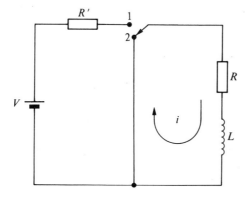

Fig. 5.9.

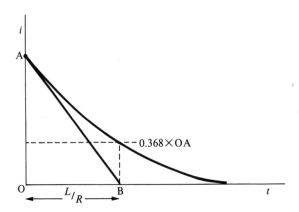

Fig. 5.10.

L/R seconds. In fact, it falls to 0.368 of its original value in this time. (Note that $0.368 + 0.632 = 1$.)

The equation giving the current i at a time t is

(5.2) $$i = I \exp\left(-\frac{Rt}{L}\right)$$

where I is the original current.

Examples 5.1

1. A coil has an inductance of 5 H and a resistance of 2 Ω. Determine its time constant.

Solution

$$\text{Time constant} = \frac{L}{R}$$
$$= \frac{5}{2}$$
$$= 2.5 \text{ s}.$$

2. The above coil is suddenly connected (at $t = 0$) to a 15 V battery. Find the final current.

Solution

$$I = \frac{V}{R}$$
$$= \frac{15}{2}$$
$$= 7.5 \text{ A}.$$

L-R circuits

3. What is the current (a) after one time constant and (b) after five time constants?

Solution

(a) $0.632 \times I$
$= 0.632 \times 7.5$
$= 4.74$ A.

(b) 0.993×7.5
$= 7.448$ A.

4. What is the current after (a) 1.25 s and (b) after 6 s.

Solution

(a) $i = 7.5 \left[1 - \exp\left(-\dfrac{1.25}{2.5}\right)\right]$
$= 7.5 \, [1 - \exp(-0.5)]$
$= 7.5 \, (1 - 0.607)$
$= 7.5 \times 0.393$
$= 2.95$ A.

(b) $i = 7.5 \left[1 - \exp\left(-\dfrac{6}{2.5}\right)\right]$
$= 7.5 \, [1 - \exp(-2.4)]$
$= 7.5 \, (1 - 0.091)$
$= 7.5 \times 0.909$
$= 6.82$ A.

5. A short circuit is placed across the coil of the previous problems when the current flowing through it is 7.5 A. Find the current after 2.5 s.

Solution

$i = 7.5 \times 0.368$
$= 2.76$ A.

6. A coil of inductance 2 H and resistance 1 Ω is connected to a 100 V d.c. supply. Find the current after 5 s and the final current. When the final current is flowing it is disconnected from the supply and switched across a 3 Ω resistor. Find the current after 1 s.

Solution

Time constant $= \tfrac{2}{1}$ s
$= 2$ s.

Current after 5 s $= \dfrac{100}{1} [1 - \exp(-\tfrac{2}{5})]$

$= 100 \, (1 - 0.67)$

$= \dfrac{100}{1} [1 - \exp(-\tfrac{2}{5})]$
$= 100 \, (1 - 0.67)$
$= 100 \times 0.33$

$$= 33 \text{ A}.$$

$$\text{Final current} = \frac{100}{1} = 100 \text{ A}.$$

When 'discharging' through the 3 Ω resistor the total resistance is 4 Ω so that the time constant is $\frac{2}{4} = 0.5$ s.

$$\text{Current after 1 s} = 100 \exp(-2)$$
$$= 100 \times 0.135$$
$$= 13.5 \text{ A}.$$

5.2 C-R circuits

Charging a capacitor

In electrical engineering we are frequently concerned with charging up a capacitor; this is often because the output of a circuit, such as an amplifier, has a *stray capacitance* which we must charge if we change the output voltage. As all circuits possess resistance, we need to consider what happens when we charge a capacitor via a resistor (Fig. 5.11). We will assume that before S is closed the capacitor has no charge. Hence, at the instant of closing the switch, there is zero volts across the capacitor. The battery voltage (V) is across the resistor (R) and the current flowing (i) must be V/R. This current flows into the capacitor, charging the top plate positively. This decreases the voltage across R to V minus the capacitor voltage. Hence the current in R decreases and we have an exponential rise of voltage across the capacitor (Fig. 5.12). Eventually the capacitor becomes charged to the voltage V (theoretically it never *quite* reaches V).

Consider point A on the curve, after a time t when the voltage across the capacitor is v_C. The voltage across R must be $(V - v_C)$ so that

$$iR = V - v_C$$

or

$$i = \frac{V - v_C}{R}$$

If this current were to remain constant until the capacitor were charged

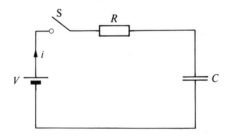

Fig. 5.11.

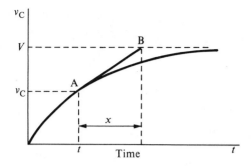

Fig. 5.12.

(line AB), taking x seconds, the quantity of electricity passing into the capacitor would be

$$ix = \frac{V - v_C}{R} x \text{ coulombs.}$$

But charge is capacitance × voltage, and the charge added in the time x is

$$(V - v_C) C$$

$$\therefore \quad \frac{V - v_C}{R} x = (V - v_C) C$$

$$x = CR.$$

This is clearly the time constant of the circuit (T) and is defined either as the time taken to charge the capacitor if the original rate of increase of voltage were to be maintained *or* the time taken for the voltage to reach 0.632 of its final value (Fig. 5.13). It will be clear that during this charging

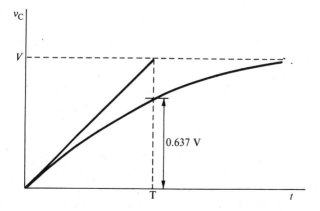

Fig. 5.13.

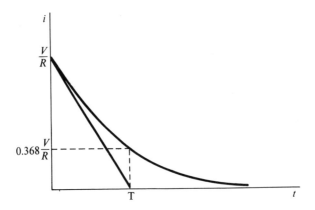

Fig. 5.14.

the current is falling from V/R to zero, when the capacitor is fully charged. It falls exponentially, with a time constant of CR (Fig. 5.14).

The actual equation for the voltage across the capacitor, v_C, after a time t is similar to eqn (5.1) for the rise in current in an L-R circuit.

(5.3) $$v_C = V\left[1 - \exp\left(-\frac{t}{CR}\right)\right].$$

The current is given by $(V - v_C)/R$ which, from eqn (5.3), is

(5.4) $$i = \frac{V - v_C}{R} = \frac{V}{R}\exp\left(-\frac{t}{CR}\right).$$

It should be noted that if we wish to make the rise of voltage across the capacitor linear, we can do so by charging it with a constant current. This is the principle of the production of saw-tooth waveforms, such as are used in the time-bases of cathode-ray oscilloscopes.

Discharging capacitors

Suppose a capacitor C charged to a voltage V is allowed to discharge through a resistor R. The voltage will fall exponentially (Fig. 5.15), with a time constant of CR. The current, which starts at V/R, will also fall exponentially as in Fig. 5.14. The equation for the fall in voltage is

(5.5) $$v_C = V\exp\left(-\frac{t}{CR}\right).$$

the current equation being the same as in the case of charging (eqn (5.4)). As all circuits must possess some capacitance, a limit is set in any circuit to the rate at which voltages can change. Whilst this is often of no practical importance, it becomes very significant in some elctronic applications where we are trying to produce instantaneous changes of voltage (e.g. in a square waveform).

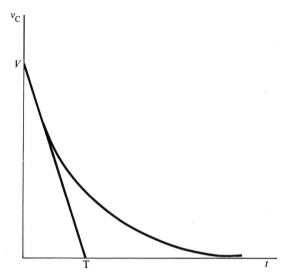

Fig. 5.15.

5A Test questions

1. Currents in a resistive d.c. circuit assume their final values instantaneously
    (a) true,
    (b) false.
    T

2. The effect of inductance in a d.c. circuit is that the current cannot change instantaneously
    (a) true,
    (b) false.
    T

3. Nearly all circuits possess self inductance
    (a) true,
    (b) false.
    F

4. All circuits include some capacitance
    (a) true,
    (b) false.
    T

5. The effect of capacitance in a d.c. circuit is that current cannot change instantaneously
    (a) true,
    (b) false.
    T

6. The effect of capacitance in a d.c. circuit is that voltages cannot change instantaneously
    (a) true,
    (b) false.
    F.

7. A coil of inductance 2 H and resistance 5 Ω is connected to a 20 V d.c. supply. The 'final' current is

(a) 10 A,
(b) 4 A,
(c) 40 A,
(d) 100 A.

8. The time constant of the coil of question 7 is
   (a) 10 s,
   (b) 0.4 s,
   (c) 2.5 s,
   (d) 0.1 s.

9. The current in the coil of question 7 reaches its final value in
   (a) 1 time constant,
   (b) 2 time constants,
   (c) 5 time constants,
   (d) never.

10. Find the current in the coil of question 7 one second after connection to the 20 V supply.

11. The current in the coil of question 7 is 4 A. It is now switched across a 3 Ω resistor. Determine
    (a) the time constant of the circuit,
    (b) the current after 0.125 s,
    (c) the 'final' current.

12. A coil of inductance 2 H and resistance 200 Ω is suddenly switched across a 100 V supply. Find
    (a) the time constant of the coil,
    (b) the final current,
    (c) the current after 20 ms,
    (d) the initial rate of increase of current.

13. A circuit has an inductance of 0.5 H and a resistance of 20 Ω. A pulse of voltage is applied such that the voltage rises instantaneously from zero to 30 V and stays at 30 V for 0.05 s before falling instantaneously back to zero. Find the maximum value reached by the current and its value 0.01 s after the 30 V has fallen to zero.

14. A coil of inductance 5 H and resistance 5 Ω is suddenly connected across a 100 V d.c. supply. Calculate the current at 0.25 s intervals for the first two seconds. Hence sketch a graph of current against time.
    When the final current (20 A) has been reached the voltage is suddenly removed and the coil switched across a 5 Ω resistor. Again calculate the current at 0.25 s intervals for the first two seconds and sketch a graph of current against time.

15. A capacitor of 2 μF is charged via a 100 MΩ resistor from a 250 V d.c. supply. Find
    (a) the time constant,
    (b) the voltage across the capacitor after 4 minutes,
    (c) the initial current,
    (d) the current after 4 minutes.

16. A capacitor of 1 μF is connected in series with a 1 MΩ resistor to a 10 V d.c. supply. Find

(a) the time constant,
(b) the initial current,
(c) the initial rate of change of current,
(d) the initial rate of change of the capacitor voltage,
(e) the capacitor voltage after 0.5 s,
(f) the current after 0.5 s.

17. A capacitor of $2\,\mu\text{F}$ stores a charge of $100\,\mu\text{C}$. It is suddenly discharged through a $500\,\text{k}\Omega$ resistor. Find
(a) the initial current,
(b) the current after 1.5 s,
(c) the capacitor voltage after 1.5 s.

18. A pulse of voltage is such that the level changes from zero to 10 V instantaneously, remains at 10 V for 1 ms and then returns instantaneously to zero. It is applied to a $100\,\text{k}\Omega$ resistor in series with a capacitor. Find the maximum voltage reached across the capacitor if its value is
(a) 10 pF,
(b) $0.02\,\mu\text{F}$.
Sketch the capacitor voltage in each case.

6 A.C. circuits I

6.1 Alternating quantities

An *alternating quantity* is one that repeatedly reverses in direction. Thus the waveforms shown in Fig. 6.1 are all alternating except for (e) which is varying but not changing direction. A *cycle* is one complete set of changes and each cycle must contain a positive part and a negative part, not necessarily equal in time, although they are in Fig. 6.1 (a)–(d). The time taken for the quantity to undergo one complete cycle is called the *periodic time*, or more often the *period*, of the waveform and is given the symbol T.

A quantity closely related to period is the *frequency*. This is the number of complete cycles which the quantity undergoes in one second. It is given the symbol f and a waveform which undergoes one cycle in one second is said to have a frequency of one *hertz* (Hz). It should be clear that

(6.1) $$f = \frac{1}{T} \quad \text{or} \quad T = \frac{1}{f}$$

so that, for example, the period of the British mains voltage waveform (50 Hz) is

$$T = \tfrac{1}{50}\,\text{s}$$
$$= 20\,\text{ms}.$$

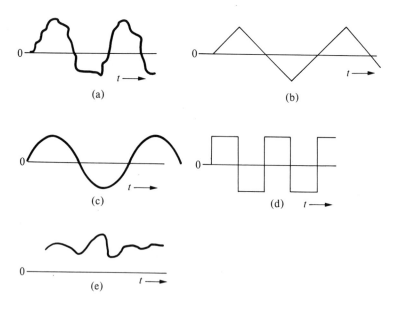

Fig. 6.1.

Similarly the frequency of a waveform which has a period of 0.5 μs is

$$f = \frac{1}{0.5 \times 10^{-6}}$$

$$= 2 \times 10^6 \, \text{Hz or 2 MHz}.$$

Alternating waveforms are encountered throughout electrical engineering. In audio amplifiers, for example, the alternating sound waves impinging on a microphone are converted into alternating voltage or current waveforms which are then amplified. In communication engineering, information is far more easily conveyed using alternating rather than unidirectional waveforms. In power engineering most of the world's supply systems use alternating rather than direct voltages. Switching problems are easier to solve and the use of a.c. in the home and in industry is safer than d.c. Very simple a.c. motors are used in such things as vacuum cleaners and spin driers. Lastly, but very importantly, a.c. can be stepped up or down in voltage by the use of transformers (Chapter 8). Thus power can be transmitted at a very high voltage and low current, using far less copper (an important economic consideration) than a low-voltage, high-current, system. The high voltage can then be transformed down to a safe value for actual use.

Sine waves

The waveform used in our power system is the *sine wave*. A sine wave can be constructed by plotting the sine of an angle, θ, against the angle itself. This is done over a complete 360° in Fig. 6.2. Note that θ has been given in radians, a much more useful angular measure in engineering. (You may remember that 360° corresponds to 2π radians.) The sine wave is the shape of the waveform produced if a coil rotates in a uniform magnetic field (Fig. 6.3). It is thus a relatively easy waveform to generate. It is also the only waveform whose shape is not changed when it is passed through a transformer—a very useful property.

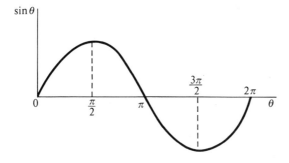

Fig. 6.2.

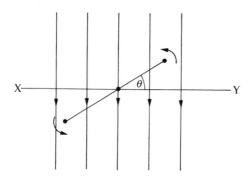

Fig. 6.3.

The waveform produced by the rotating coil of Fig. 6.3 is

(6.2) $$v = V_{max} \sin \theta$$

where V_{max} is the maximum value attained by v during the cycle and θ is the angle which the coil is making at any instant with the line XY shown in Fig. 6.3. v is the value at a particular instant and is called the *instantaneous value*.

If the coil rotates at f rotations per second, f will be the frequency of the waveform produced. The angular velocity of the coil, that is the number of radians turned through per second, is obviously $2\pi f$ and this is given the symbol ω (omega). Thus the angle turned through in a time t seconds is ωt or $2\pi f t$ and eqn (6.2) becomes

$$v = V_{max} \sin \omega t$$
$$= V_{max} \sin 2\pi f t.$$

The maximum value of v is more often referred to as the *peak value*, V_p, and this is the symbol which will be used. Thus

(6.3) $$v = V_p \sin \omega t$$
$$= V_p \sin 2\pi f t.$$

A quantity often quoted, particularly in electronics, is the *peak-to-peak value* (Fig. 6.4). For many waveforms, including the sine wave, it is twice the peak value. Great care must be taken not to confuse the two. Note that the time taken to complete one cycle is $2\pi/\omega$ that is

(6.4) $$T = \frac{2\pi}{\omega}.$$

It can be seen that as $\omega = 2\pi f$

$$T = \frac{2\pi}{2\pi f} = \frac{1}{f}$$

as we already know (eqn (6.1)).

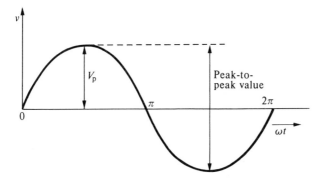

Fig. 6.4.

Eqn (6.3) can clearly be used to calculate the instantaneous value of a sine wave. For example, if we need to find the instantaneous value of a sinusoidal voltage of peak value 20 V and frequency 150 Hz, 1 ms after passing through zero and increasing, it can be done using this equation:

$$v = V_p \sin \omega t.$$

Now $\omega = 2\pi f$
$= 2\pi \, 150,$

hence $\omega t = 2\pi \, 150 \times 10^{-3}$
$= 0.942 \text{ rad}$
$= \dfrac{0.942 \times 180}{\pi} \,°$
$= 53.97°$

$v = 20 \sin 53.97$
$= 16.17 \, V.$

Instantaneous values are of little use in most electrical engineering problems. The peak value is much more important, as are two other values defined below.

Average value

The *average* seems a very obvious way in which to measure an alternating quantity. However, many waveforms have identical positive and negative half cycles and the average would thus be zero, a value with little meaning in this context. However, it is very useful to know the average value of one of the half cycles, and this is, by definition, what we mean when we talk of the average value of an alternating quantity. It is particularly useful in rectification circuits, as you will find in your later study of electronics.

The actual average value depends, of course, on the shape of the waveform. For a sine wave it can be shown that

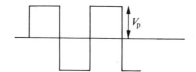

Fig. 6.5.

(6.5) $$V_{AV} = \frac{2V_p}{\pi} = 0.637 V_p.$$

It must be emphasized that this result applies *only* to a sine wave. For example, Fig. 6.5 shows a square wave, a common waveform in electronic engineering. Obviously, the average over a half cycle is simply V_p, that is the average and peak values are equal for a *square* wave.

R.M.S. value

In most engineering applications the most important thing about a waveform is the power that it can produce. Now we know that if a direct current of I flows in a resistance R it produces a power I^2R. The direction of the current is immaterial. An alternating current will also produce power in a resistor. It will do so on both half cycles and the r.m.s. value of an alternating current may be defined as the value of the direct current that would produce the same power in the same resistance. As power is proportional to I^2 we need to find the mean of the squares of the instantaneous values. The r.m.s. value will be the square root of this and hence the name; <u>r</u>oot <u>m</u>ean <u>s</u>quare. It is conventional, when talking about a voltage or a current, to quote the r.m.s. value unless otherwise stated. Thus the mains voltage of 240 V is actually 240 V r.m.s.

For a sine wave it can be shown that

(6.6) $$V_{r.m.s.} = \frac{V_p}{\sqrt{2}} = 0.707 V_p.$$

Again it must be emphasized that these figures only apply to a sine wave.

The square wave of Fig. 6.5 has an r.m.s. value of V_p. The square wave is unique in that the r.m.s., average, and peak values are all equal.

Form factor

The form factor of an alternating quantity is defined as

(6.7) $$\text{form factor} = \frac{\text{r.m.s. value}}{\text{average value}}.$$

The importance of form factor is that it gives the engineer an idea of the shape of the waveform and it can easily be obtained in practice, by measuring the r.m.s. and average values. For a sine wave

$$\text{form factor} = \frac{0.707 V_p}{0.637 V_p}$$

$$= 1.11.$$

Similarly, for a square wave

$$\text{form factor} = \frac{V_p}{V_p}$$

$$= 1.$$

In general, if the form factor is less than 1.11 the waveform is becoming squarer than a sine wave, and if it is greater than 1.11 it is becoming more triangular.

6.2 Phasors

Phase

For the rest of this chapter and in Chapter 7 we shall be only dealing with sine waves. Any repeating waveform can be made by adding together a number of sine waves, so that the effect of circuits on sinusoidal quantities is of paramount importance.

A single sine wave can be specified by two quantities, frequency (or period) and magnitude. The magnitude can be quoted as peak, peak-to-peak, average, or r.m.s. Because these are all related for a sine wave, a knowledge of any one enables the others to be calculated.

There is another characteristic which must be considered when dealing with more than one sine wave. This is the time at which it passes through zero. The two sine waves of Fig. 6.6 are both passing through zero and increasing together and are said to be *in phase*. They are both shown as voltages, although they could both be currents or one a voltage and the

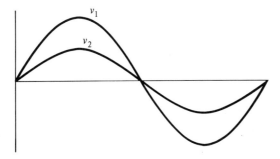

Fig. 6.6.

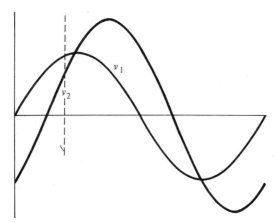

Fig. 6.7.

other a current. Two sine waves are not, of necessity, in phase; Fig. 6.7 shows two of the same frequency which are not in phase. At any instant, such as that shown by the vertical dotted line, v_1 is further on in its cycle than v_2. v_1 is ahead of v_2. It is said to be *leading* v_2. The amount by which one waveform is *leading* or *lagging* another is measured in degrees or radians, taking, as usual, 360° or 2π radians as a complete cycle. In Fig. 6.7, v_1 leads v_2 by about 45° or $\pi/4$ rad. One could, of course, say that v_2 leads v_1 by 315° but it is conventional to quote the smaller angle.

Fig. 6.8 shows two waveforms 180° or π rad out of phase. v_1 is leading or lagging v_2 by π rad.

Addition of sine waves

We shall often be concerned with adding two or more sine waves. v_1 and v_2 in Fig. 6.7 might be the output voltage waveforms of two sine-wave

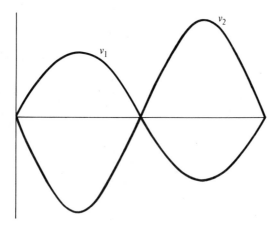

Fig. 6.8.

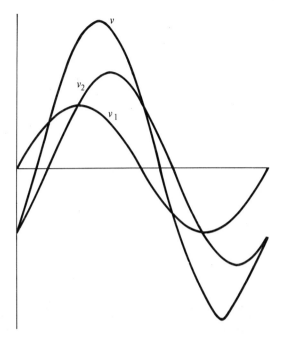

Fig. 6.9.

generators connected in series and we might need to find v, the output of the combination. Similarly they might be the waveforms of two currents flowing into a common wire, and thus adding. The addition can be done graphically, as shown in Fig. 6.9, by simply adding instantaneous values at as many points on the cycle as is convenient. It is found that v is also a sine wave of the same frequency as v_1 and v_2, and somewhere between v_1 and v_2 in phase. In the example shown it is larger in magnitude than either v_1 or v_2 but this is not necessarily so, as can be seen in Fig. 6.10 where v_2

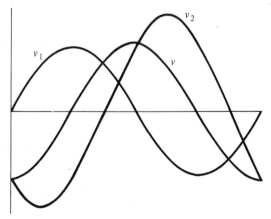

Fig. 6.10.

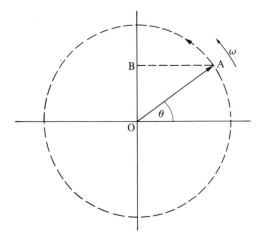

Fig. 6.11.

is lagging v_1 by about 135°. Here v is less than v_2. We must get used to the fact that two out of phase alternating waveforms do not add arithmetically and the resultant can be less than either or both of the waveforms. In fact, two waveforms of equal magnitude but 180° out of phase have a resultant of zero!

Clearly the above procedure is not a very convenient method of adding sine waves, and other methods are available. The method to be considered here is the representation of the sine wave by a *phasor*. The sine wave is represented by a line (OA) carrying an arrowhead, which rotates, anti-clockwise by convention, about the point O as shown in Fig. 6.11. It is imagined to be rotating at an angular velocity ω where

$$\omega = 2\pi f.$$

If OA represents the maximum value, V_p, of the quantity and the angle θ the angle which the line OA makes with the zero axis, then the projection of OA on the vertical, OB, is given by

$$OB = OA \sin \theta$$
$$= V_p \sin \theta,$$

or, as $\theta = \omega t$

(6.8) $$OB = V_p \sin \omega t.$$

Comparing this with eqn (6.3) shows that OB represents the instantaneous value of the waveform. It is zero when θ is 0 or 180° and a maximum positive at 90° and maximum negative at 270° (or −90°). Now as we are not usually concerned with instantaneous values, the diagram of Fig. 6.11 (called a *phasor diagram*) is of little practical value. However, if both waveforms of, say, Fig. 6.9 are drawn on the same diagram we get Fig. 6.12. Note that v_2 is larger than v_1 (because the magnitude of v_2 is greater than

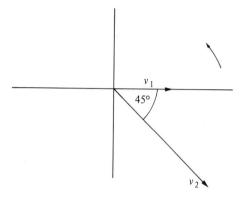

Fig. 6.12.

v_1) and that v_1 is leading v_2 by 45°. This diagram thus shows the two main differences between the two waveforms. Of course, the phasors could be drawn anywhere on their rotation, but v_1 must be 45° ahead of v_2 and smaller in size. Fig. 6.13 shows a similar diagram. In general Fig. 6.13 gives no useful information which is not contained in Fig. 6.12. It is usually convenient to draw one phasor along the zero axis.

A glance at Fig. 6.12 shows the phase and magnitude differences between the waveforms much more quickly than Fig. 6.9. Even more important, it enables the sum of v_1 and v_2 to be easily found. This is done by completing the parallelogram of which v_1 and v_2 form two sides. The resultant, v, is then the diagonal, as shown in Fig. 6.14. This is the same procedure as used in the parallelogram of forces in mechanics for finding the resultant of two forces.

The length v can be measured, as can ϕ, the angle by which it (v) is lagging v_1. Obviously it leads v_2 by $(45 - \phi)$.

These three voltages can be represented algebraically. If

$$v_1 = V_{1p} \sin \omega t$$

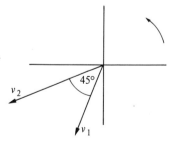

Fig. 6.13.

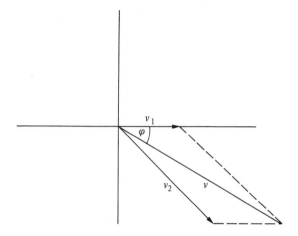

Fig. 6.14.

then, because v_2 lags v_1 by $\pi/4$ radians it is written

$$v_2 = V_{2p} \sin(\omega t - \pi/4).$$

Whatever the value of ωt, the angle of v_2 is $\pi/4$ radians behind it. It is more correct to specify the angle of lead or lag in radians because ωt is in radians. Similarly

$$v = V_p \sin(\omega t - \phi).$$

Fig. 6.15 shows a sine wave, v_1, of amplitude 50 V, 60° ahead of one, v_2, of amplitude 30 V. The resultant, v, can be seen to have an amplitude of about 70 V and to lag v_1 by an angle ϕ of about 20°.

Scale drawings of this type are very time consuming, and the solution can be found mathematically by working in horizontal and vertical components. A phasor can be resolved into two components at right angles, as in Fig. 6.16. OA shows the phasor v_2, of magnitude 30 V and angle $-60°$ with the horizontal. It is negative because positive angles are, by convention, taken as anticlockwise. The components of it are OC (horizontal) and OB (vertical). Obviously

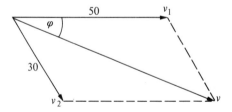

Fig. 6.15.

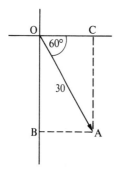

Fig. 6.16.

$$OC = OA \cos(-60)$$
$$= 30 \cos 60 \text{ as } \cos(-\theta) = \cos\theta$$
$$= 15\,\text{V},$$

and $OB = OA \sin(-60)$
$$= -30 \sin 60 \text{ as } \sin(-\theta) = -\sin\theta$$
$$= -26\,\text{V},$$

so that OA can be specified by a horizontal phasor of magnitude 15 V and a vertical one of magnitude 26 V (acting downwards). v_1 is horizontal, of magnitude 50 V, and so has no vertical component. The *total* horizontal component is

$$50 + 15 = 65\,\text{V},$$

and the *total* vertical one is -26 V. Thus the resultant, v, has components as shown in Fig. 6.17. Its magnitude is

$$\sqrt{(65^2 + 26^2)} = 70\,\text{V}$$

and ϕ is given by

$$\phi = \tan^{-1} \tfrac{26}{65}$$
$$= 21.8°\text{ lagging.}$$

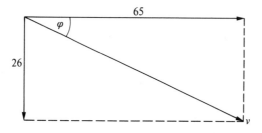

Fig. 6.17.

102 A.C. circuits I

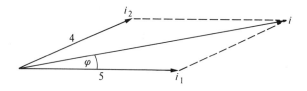

Fig. 6.18.

We have assumed that the phasors referred to above represent the peak or maximum values. As phasors are only used to represent sine waves, and as any sine wave has the same relationship between r.m.s. and peak, and between average and peak, the phasors could just as well be representing r.m.s. or average values. Of course, all the phasors on one diagram must be drawn to represent the same measurement.

Examples 6.1

1. Two currents i_1 and i_2 have r.m.s. values of 5 A and 4 A respectively. If i_2 leads i_1 by 30° find, by scale drawing, the sum of i_1 and i_2.

Solution

See Fig. 6.18.
   By measurement

$$i = 8.7 \text{ A},$$
$$\phi = 13° \text{ leading } i_1.$$

2. Repeat question 1 by calculation.

$$i_1 \text{ horizontal} = 5 \text{ A},$$
$$i_1 \text{ vertical} = 0,$$
$$i_2 \text{ horizontal} = 4 \cos 30 = 3.46 \text{ A},$$
$$i_2 \text{ vertical} = 4 \sin 30 = 2.00 \text{ A},$$

$$\text{total horizontal} = 5 + 3.46 = 8.46 \text{ A},$$
$$\text{total vertical} = 2.00 \text{ A},$$

$$\text{magnitude of } i = \sqrt{(8.46^2 + 2^2)} = 8.69 \text{ A},$$
$$\phi = \tan^{-1} \frac{2}{8.46} = 13.3° \text{ leading } i_1.$$

3. Two voltages v_1 and v_2 are represented by

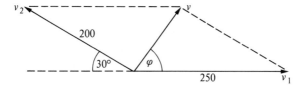

Fig. 6.19.

$$v_1 = 250 \sin \omega t \text{ V}$$
$$v_2 = 200 \sin (\omega t + 5\pi/6) \text{ V}.$$

Draw a phasor diagram to represent them, and find the sum

$$v = v_1 + v_2$$

expressing the answer as above.

Solution

See Fig. 6.19.

$$5\pi/6 \text{ rad} \equiv 150°,$$
$$v_1 \text{ horizontal} = 250 \text{ V},$$
$$v_1 \text{ vertical} = 0,$$
$$v_2 \text{ horizontal} = -200 \cos 30 = -173.2 \text{ V},$$
$$v_2 \text{ vertical} = 200 \sin 30 = 100 \text{ V},$$
$$\text{total horizontal} = 250 - 173.2 = 76.8 \text{ V},$$
$$\text{total vertical} = 100 \text{ V},$$
$$\text{magnitude of } v = \sqrt{(76.8^2 + 100^2)} = 126 \text{ V},$$
$$\phi = \tan^{-1} \frac{100}{76.8} = 52.5° \text{ or } 0.92 \text{ rad},$$
$$v = 126 \sin (\omega t + 0.92).$$

4. Fig. 6.20 shows the currents in three wires. If

$$i_1 = 5 \sin (\omega t + \pi/6)$$
$$i_3 = 10 \sin (\omega t + \pi/3)$$

show i_1 and i_3 on a phasor diagram and hence find i_2.

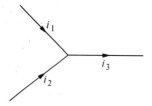

Fig. 6.20.

Solution

See Fig. 6.21.
By Kirchhoff's Law

$$i_3 = i_1 + i_2$$

hence $\quad i_2 = i_3 - i_1,$

horizontal component of i_3 = 10 cos 60 = 5 A,
horizontal component of i_1 = 5 cos 30 = 4.33 A,

hence horizontal component of i_2 = 5 − 4.33 = 0.67 A,

vertical component of i_3 = 10 sin 60 = 8.66 A,
vertical component of i_1 = 5 sin 30 = 2.5 A,

hence vertical component of i_2 = 8.66 − 2.5 = 6.16 A,

magnitude of $i_2 = \sqrt{(0.67^2 + 6.16^2)} = 6.2$ A,

$$\phi = \tan^{-1} \frac{6.16}{0.67} = 83.8° \text{ or } 1.46 \text{ rad},$$

hence $\quad\quad\quad\quad\quad\quad i_2 = 6.2 \sin(\omega t + 1.46).$

Note that i_3 will form the diagonal of the parallelogram which has i_1 and i_2 as two of its sides.

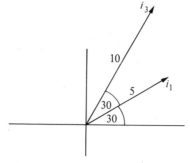

Fig. 6.21.

6A Test questions

1. Which of the waveforms shown in Fig. 6.22 are alternating?

2. A waveform has a frequency of 10 MHz. Its periodic time is
   (a) 10 μs,
   (b) 0.1 μs,
   (c) 0.1 s,
   (d) 10 ms.

3. A sinusoidal voltage is represented by

   $200 \sin \omega t$ V.

   If the frequency is 100 Hz find the instantaneous voltage after
   (a) 1.25 ms,
   (b) 5.5 ms.

4. A sinusoidal voltage has a frequency of 60 Hz and an r.m.s. value of 200 V. Write an equation for the instantaneous value, assuming that it is zero (increasing) at $t = 0$, and find this value (a) 2.5 ms and (b) 12.5 ms after passing through its maximum positive value.

5. At what time after passing through zero will the waveform of question 4 have an instantaneous value of 141.4 V?

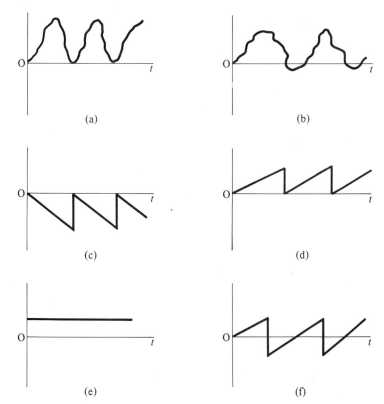

Fig. 6.22.

106   A.C. circuits I

6. Determine the average and r.m.s. values of the waveform of question 3.

7. A waveform has an average value of 25 and an r.m.s. value of 27. What is its form factor?

8. Two voltages, V_1 and V_2, have r.m.s. values of 200 V and 240 V respectively. V_1 leads V_2 by 45°. Draw a scale phasor diagram and hence determine the sum of V_1 and V_2 in magnitude and phase relative to V_1.

9. Repeat question 8 by calculation.

10. Find the phasor sum of the following four currents

$i_1 = 25 \sin \omega t$,
$i_2 = 30 \sin(\omega t + \pi/6)$,
$i_3 = 20 \sin(\omega t - \pi/4)$,
$i_4 = 30 \sin(\omega t + \pi/2)$.

11. Three voltages are represented by

$v_1 = 500 \sin \omega t$,
$v_2 = 300 \sin(\omega t + \pi/2)$,
$v_3 = 700 \sin(\omega t + \pi/6)$.

Find the r.m.s. value of the resultant and its phase relative to v_3.

12. Find the phasor sum of

$i_1 = 10 \sin \omega t$,
$i_2 = 10 \sin(\omega t + 2\pi/3)$,
$i_3 = 10 \sin(\omega t - 2\pi/3)$.

6.3 Single-phase a.c. circuits

Resistance in an a.c. circuit

If a voltage given by

$$v = V_p \sin \omega t$$

is applied to a perfect resitor, R, the instantaneous current which flows is clearly given by

$$i = \frac{v}{R}$$

(6.9)   hence $i = \frac{V_p}{R} \sin \omega t$.

This is the equation of a sine wave in phase with the voltage and of peak value

(6.10) $$I_p = \frac{V_p}{R}.$$

(6.11) Similarly $$I_{AV} = \frac{V_{AV}}{R}$$

(6.12) and $$I_{r.m.s.} = \frac{V_{r.m.s.}}{R}.$$

Hence calculations involving resistance in an a.c. circuit are very similar to those involving resistance in a d.c. circuit. For example if a voltage of 240 V r.m.s. is applied to a 10 Ω resistor the current flowing is

$$\frac{240}{10} = 24 \text{ A r.m.s.}$$

By the definition of r.m.s. values given earlier, the power dissipated in the resistor is

(6.13) $$P = I_{r.m.s.}^2 R$$

and in the above example is

$$24^2 \times 10 = 5760 \text{ W}.$$

The power could also have been calculated using

$$P = \frac{V^2}{R} = IV.$$

Inductance in an a.c. circuit

Although there is no such thing as a perfect inductor, it is useful to consider such a component and then later to see the effect of taking into account its resistance. We saw in Chapter 4 that if a current flows in an inductor the voltage across it is proportional to the rate at which the current is changing. It is, in fact, equal to the rate of change of current times the inductance, L (eqn (4.14)). If we pass a sinusoidal current through a perfect inductor we shall produce a voltage across it proportional to the rate at which the current changes. Now a sine wave has a maximum rate of change when its value is passing through zero, and a rate of change of zero when it is at its maximum values (positive and negative). Fig. 6.23 shows a sinusoidal current, i, and a graph, the dotted line, showing the rate of change of the current:

> At time A the rate of change of current is maximum positive.
> At time B the rate of change of current is zero.
> At time C the rate of change of current is maximum negative.
> At time D the rate of change of current is zero, etc.

The waveform, the dotted line, will be the waveform of the voltage across the inductor. As can be seen it is 90° or $\pi/2$ rad ahead of the current. Current lags voltage in an inductor.

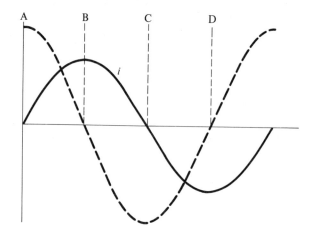

Fig. 6.23.

The magnitude of the voltage depends on three factors: (i) the magnitude of the current, (ii) the maximum rate of change of current, which is clearly dependant on the frequency or ω, and (iii) the inductance, L. It is, in fact, proportional to all three, and the peak voltage is given by

(6.14) $$V_p = \omega L I_p.$$

If the current waveform is

$$i = I_p \sin \omega t$$

the voltage waveform will be

(6.15) $$\begin{aligned} v &= V_p \sin(\omega t + \pi/2) \\ &= \omega L I_p \sin(\omega t + \pi/2). \end{aligned}$$

As, for any sinusoidal quantity, the r.m.s. value is equal to the peak divided by $\sqrt{2}$, eqn (6.14) can be written

(6.16) $$V = \omega L I$$

where V and I are r.m.s. quantities.

These quantities are shown as a phasor diagram in Fig. 6.24, each, of course, being drawn to its own scale.

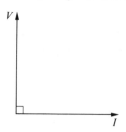

Fig. 6.24.

In a resistive circuit the ratio V/I is the opposition to a flow of current, and is called the resistance of the circuit. Clearly V/I is still the opposition to current flow in an inductor and, from eqn (6.16)

(6.17) $$\frac{V}{I} = \omega L.$$

It is not the same thing as resistance, as will be seen below, and it is called the reactance of the inductor. The symbol for inductive reactance is X_L, so that

(6.18) $$X_L = \omega L = 2\pi f L.$$

As X_L is a voltage divided by a current, the unit used is the ohm. It is very important to note that inductive reactance is proportional to both the inductance and the frequency. It is zero at zero frequency (d.c.) and infinite at an infinite frequency.

Thus a coil of inductance 1 H at 50 Hz has a reactance of

$$X_L = 2\pi\, 50 \times 1 = 314\,\Omega$$

and it if it connected to the mains (240 V) the current flowing is

$$I = \tfrac{240}{314} = 0.76\,\text{A}.$$

Power in an inductor

At any instant power in a circuit is the product of voltage and current. Thus the power in an inductor can be found by multiplying the instantaneous values of voltage and current on the graph of Fig. 6.23. This has been done in Fig. 6.25 over a complete cycle. It should be observed that

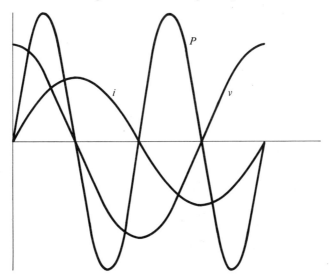

Fig. 6.25.

110 A.C. circuits I

the power is zero when either v or i is zero and is negative when either v or i, but not both, is negative. The power waveform is a sinusoidal wave of twice the frequency of the others. Power is put into and taken out of the inductor twice during each cycle of the supply. The inductor stores energy (Chapter 4) and then returns it. The average power dissipated is zero. A pure inductor cannot dissipate power. It can only store it and return it to the supply. This means that our 1 H inductor connected to the mains, although taking a current of 0.76 A, absorbs no power and hence costs us nothing to run. Of course, because it takes no power it will do no work and is not of much use in this context.

Capacitance in an a.c. circuit

A perfect capacitor connected to an a.c. supply behaves in a somewhat different way to an inductor. Current flows into or out of a capacitor to alter the voltage across its plates. It will only flow if the voltage is changing and will be proportional to the rate of change, that is to ω. If a sinusoidal voltage is applied (Fig. 6.26) the current will be as shown. The current is 90° ahead of the voltage. Current leads voltage in a capacitor. The magnitude of the current will be proportional to the amplitude of the voltage, the rate at which the voltage is changing, and the capacitance C because a larger capacitance requires a larger current to change its voltage at a given rate. In fact

(6.19) $$I_p = \omega C V_p.$$

If the voltage waveform is

$$v = V_p \sin \omega t$$

the current waveform will be

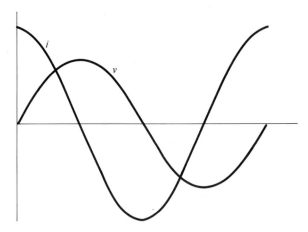

Fig. 6.26.

(6.20) $$i = I_p \sin(\omega t + \pi/2)$$
$$= \omega C V_p \sin(\omega t + \pi/2).$$

Using r.m.s. values

$$I = \omega C V$$

(6.21) or $V = \dfrac{I}{\omega C}.$

Fig. 6.27 shows a phasor diagram of voltage and current in a capacitor. The ratio V/I is again called reactance, this time the capacitvie reactance X_C and

(6.22) $$X_C = \frac{V}{I} = \frac{1}{\omega C}.$$

Note that capacitive reactance is inversely proportional to both capacitance and frequency. It is infinite at zero frequency (d.c.), that is a capacitor will pass no current at d.c., as we already know.

A capacitor of 1 µF has, at 50 Hz, a reactance of

$$X_C = \frac{1}{2\pi \times 50 \times 10^{-6}}$$
$$= 3183 \,\Omega.$$

Connected to the mains, the current will be

$$I = \frac{V}{X_C} = \frac{240}{3183} = 0.075 \text{ A}.$$

The power waveform can be drawn in a manner similar to that of Fig. 6.25, the average power again being zero. Like the inductor, the capacitor stores energy during part of each cycle, returning it at other times. No power is dissipated in a capacitor.

Examples 6.2

1. A resistance of 10 kΩ is connected to a sinusoidal voltage of 30 V, 20 kHz. Find the current and average power.

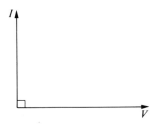

Fig. 6.27.

Solution

$$I = \tfrac{30}{10} = 3\,\text{mA},$$

$$P = I^2 R$$
$$= (3 \times 10^{-3})^2\, 10 \times 10^3$$
$$= 0.09\,\text{W} \text{ or } 90\,\text{mW}.$$

2. The supply of question 1 is connected to an inductor of inductance 80 mH. Find the current and average power.

Solution

$$X_L = 2\pi f L$$
$$= 2\pi\, 20 \times 10^3 \times 80 \times 10^{-3}$$
$$= 10\,000\,\Omega \text{ or } 10\,\text{k}\Omega,$$

$$I = \tfrac{30}{10} = 3\,\text{mA lagging the voltage by } 90°.$$

The average power is zero.

3. A current of $10\,\mu\text{A}$ at a frequency of 5 MHz is flowing through a capacitor. If the voltage across it is 0.05 V (lagging the current by 90°) find the capacitance.

Solution

$$V = \frac{I}{\omega C}$$

$$C = \frac{I}{\omega V}$$

$$= \frac{10 \times 10^{-6}}{2\pi \times 5 \times 10^6 \times 0.05}$$

$$= 6.4 \times 10^{-12}\,\text{F} \text{ or } 6.4\,\text{pF}.$$

L-R and C-R *circuits*

The preceding paragraphs have considered a single component, R, L, or C, connected to an a.c. supply. In many circuits, however, we have to deal with components in various combinations. In this chapter we shall consider series circuits, parallel circuits being considered in the next chapter.

Consider first the case of R and L in series (Fig. 6.28). Here v is a sine-wave generator of r.m.s. value V and angular frequency ω. We know that the resistor offers an opposition to current flow given by R and that the inductor offers an opposition given by ωL. However, these quantities do not simply add, because their effect on phase must be taken into account. To determine the relationship between current and voltage a phasor diagram may be drawn (Fig. 6.29).

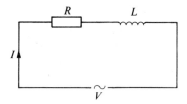

Fig. 6.28.

Firstly the phasor for current, I, is drawn. It is shown here, for convenience, along the zero axis. Now the voltage across the resistor is IR and is in phase with the current. The voltage across the inductor is $\omega L I$ and is 90° ahead of the current. These voltages must, by Kirchhoff's Second Law, add to produce the applied voltage, V. This addition is shown in the diagram, the parallelogram here being a rectangle. The magnitude of the applied voltage is found using Pythagarus

(6.23)
$$V = \sqrt{(I^2 R^2 + \omega^2 L^2 I^2)} \\ = I\sqrt{(R^2 + \omega^2 L^2.)}$$

The opposition to current flow, V/I, thus

(6.24)
$$\frac{V}{I} = \sqrt{(R^2 + \omega^2 L^2)}.$$

This opposition, in a circuit containing resistance and reactance, is called *impedance*, and has the symbol Z. Like resistance and reactance it is measured in ohsm.

The phase angle, ϕ, by which the current lags the voltage, is given by

(6.25)
$$\tan \phi = \frac{\omega L I}{R I} = \frac{\omega L}{R} = \frac{X_L}{R}$$

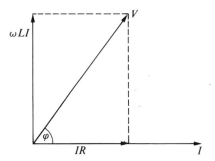

Fig. 6.29.

114   A.C. circuits I

It is also useful to note the expressions for $\sin\phi$ and $\cos\phi$

$$\sin\phi = \frac{\omega L I}{V}.$$

Now $V = IZ$,

(6.26)   hence $$\sin\phi = \frac{\omega L I}{ZI} = \frac{\omega L}{Z} = \frac{X_L}{Z},$$

(6.27)   and $$\cos\phi = \frac{RI}{V} = \frac{RI}{ZI} = \frac{R}{Z}.$$

The current I flowing will produce a power of $I^2 R$ in the resistor. As there is no average power in an inductor, the total power will be

(6.28) $$P = I^2 R.$$

Another useful expression for the power may be found as follows

$$P = I^2 R$$
$$= I \times I \times R$$
$$= \frac{V}{Z} \times I \times R.$$

But $\frac{R}{Z} = \cos\phi$ (eqn (6.27)),

(6.29)   hence $$P = V \times I \times \cos\phi.$$

Consider a coil of resistance 300 Ω and inductance 1 H. At 50 Hz

$$X_L = 2\pi \times 50 \times 1$$
$$= 314 \, \Omega.$$

The impedance will be

$$Z = \sqrt{(300^2 + 314^2)}$$
$$= 434 \, \Omega.$$

If connected to the 240 V mains the current will be

$$I = \tfrac{240}{434} = 0.55 \, \text{A}.$$

It will lag the voltage by ϕ where

$$\phi = \tan^{-1} \tfrac{314}{300} = 46.3°.$$

The power dissipated is

$$P = (0.55)^2 \, 300$$
$$= 90.8 \, \text{W}.$$

It is interesting to note the impedance at other frequencies. Hence at 1 Hz

$$X_L = 2\pi \times 1 \times 1$$
$$= 6.3 \, \Omega.$$
$$Z = \sqrt{(300^2 + 6.3^2)}$$
$$= 300.7 \, \Omega$$

with a phase angle

$$\tan^{-1} = \frac{6.3}{300} = 1.2°,$$

i.e. the circuit at this frequency is almost a pure resistor. Similarly at, say, 10 kHz

$$X_L = 2\pi \times 10 \times 10^3 \times 1$$
$$= 62\,832 \, \Omega$$
$$Z = \sqrt{(300^2 + 62\,832^2)}$$
$$= 62\,832.7 \, \Omega$$

with a phase angle

$$\phi = \tan^{-1} \frac{62\,832}{300} = 89.7°$$

that is the circuit, at this frequency, is almost a pure inductor.

Fig. 6.30 shows a resistor in series with a capacitor; the phasor diagram is shown in Fig. 6.31 with the voltage across C, equal to $V/\omega C$, 90° behind the current. The applied voltage is

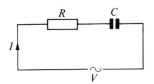

Fig. 6.30.

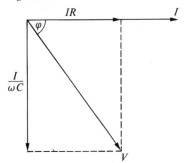

Fig. 6.31.

116 A.C. circuits I

(6.30)
$$V = I\sqrt{\left(R^2 + \frac{1}{\omega^2 C^2}\right)}$$
$$= I\sqrt{(R^2 + X_C^2)}$$

V/I is again called the impedance and

(6.31)
$$Z = \sqrt{\left(R^2 + \frac{1}{\omega^2 C^2}\right)}$$
$$= \sqrt{(R^2 + X_C^2)}.$$

The phase angle ϕ (current ahead of voltage) is given by

(6.32)
$$\tan \phi = \frac{X_C}{R} = \frac{1}{\omega C R}$$

and the power in the circuit by

(6.33)
$$P = I^2 R = VI \cos \phi.$$

Consider a capacitor of 100 pF in series with a 3 kΩ resistor connected to a 5 V, 500 kHz supply

$$X_C = \frac{1}{\omega C}$$
$$= \frac{1}{2\pi \times 500 \times 10^3 \times 100 \times 10^{-12}}$$
$$= 3183 \, \Omega$$

$$Z = \sqrt{(3000^2 + 3183^2)}$$
$$= 4374 \, \Omega \text{ or } 4.374 \, k\Omega.$$

$$I = \frac{5}{4.374} = 1.14 \, \text{mA}.$$

Impedance triangle

The triangle formed by the three phasors in the L–R circuit (Fig. 6.29) is redrawn in Fig. 6.32. If each side is divided by I, a similar triangle is

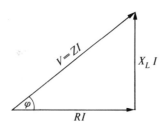

Fig. 6.32.

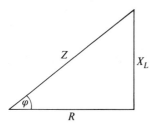

Fig. 6.33.

formed (Fig. 6.33). This has sides proportional to R, X_L, and Z and is called the *impedance triangle* of the circuit. It also gives, of course, the relationships of eqns (6.25), (6.26), and (6.27). A similar impedance triangle can be formed from the phasor diagram of the C-R circuit (Fig. 6.31).

L-C-R *circuit*

Fig. 6.34 shows a series circuit containing inductance, capacitance, and resistance. The voltages across each component, V_L, V_C, and V_R are given by:

$V_L = IX_L$ 90° ahead of the current.

$V_C = IX_C$ 90° behind the current.

$V_R = IR$ in phase with the current.

The sum of these three, added on a phasor diagram, must equal the applied voltage, V. Fig. 6.35 shows such a diagram, it being assumed here that

$$V_L > V_C.$$

Because IX_C is separated from IX_L by 180° it subtracts from it giving $I(X_L - X_C)$ as shown. The phasor sum of $I(X_L - X_C)$ and IR is V where

(6.34) $$V = I\sqrt{[R^2 + (X_L - X_C)^2]}$$

leading the current by ϕ where

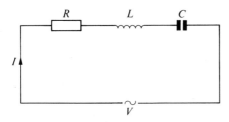

Fig. 6.34.

118   A.C. circuits I

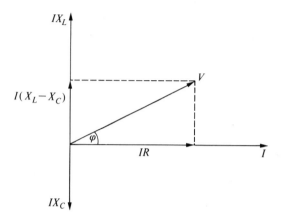

Fig. 6.35.

(6.35) $$\tan \phi = \frac{X_L - X_C}{R}.$$

Obviously

(6.36) $$Z = \frac{V}{I} = \sqrt{[R^2 + (X_L - X_C)^2]}.$$

Sometimes $(X_L - X_C)$ is simply written as X, the reactance of the circuit. Then

(6.37) $$V = I\sqrt{(R^2 + X^2)},$$

(6.38) $$\tan \phi = \frac{X}{R},$$

(6.39) and $$Z = \sqrt{(R^2 + X^2)}.$$

Fig. 6.36 shows the case when $V_C > V_L$.

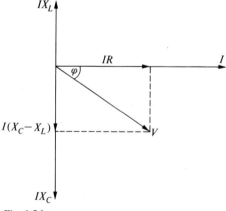

Fig. 6.36.

The equations will be the same as above. Consider a supply of 240 V, 50 Hz connected to a series circuit with

$$R = 100\ \Omega$$
$$L = 0.15\ \text{H}$$
$$C = 20\ \mu\text{F},$$
$$\begin{aligned}X_L &= 2\pi fL \\ &= 2\pi \times 50 \times 0.15 \\ &= 47.1\ \Omega,\end{aligned}$$

and $X_C = \dfrac{1}{2\pi fC}$

$$= \dfrac{1}{2\pi \times 50 \times 20 \times 10^{-6}}$$
$$= 159.2\ \Omega.$$

The circuit is capacitative because $X_C > X_L$

$$\begin{aligned}X_L - X_C &= 47.1 - 159.2 \\ &= -112.1\ \Omega \\ Z &= \sqrt{[100^2 + (-112.1)^2]} \\ &= 150.2\ \Omega. \\ I &= \dfrac{240}{159.2} = 1.6\ \text{A}. \\ \phi &= \tan^{-1}\dfrac{-112.1}{100} \\ &= -48.3\ \text{(voltage lagging current)}.\end{aligned}$$

It should be noted that the voltages across the three components are

$$V_R = IR = 1.6 \times 100 = 160\ \text{V},$$
$$V_L = IX_L = 1.6 \times 47.1 = 75.4\ \text{V},$$
$$V_C = IX_C = 1.6 \times 159.2 = 254.7\ \text{V}.$$

These do not add *arithmetically* to give 240 V, in fact the voltage across the capacitor is actually greater than the applied voltage. V_C is 180° out of phase with V_L and the voltage across the combination is $254.7 - 75.4 = 179.3$ V. This adds to V_R *taking phase into account* to give the applied voltage thus

$$\sqrt{(160^2 + 179.3^2)} = 240\ \text{V}.$$

The power in the circuit is I^2R, as no power is dissipated in the inductor or capacitor.

$$P = I^2R = 1.6^2 \times 100 = 256\ \text{W}.$$

120 A.C. circuits I

Examples 6.3

1. A coil has a resistance of 10 Ω and an inductance of 20 mH. When it is connected to a 0.5 V sinusoidal supply the current flowing is 5 mA. Find the frequency of the supply and the phase of the current relative to the voltage.

Solution

$$Z = \frac{V}{I} = \frac{0.5}{5} = 0.1 \text{ k}\Omega = 100 \, \Omega$$

$$\sqrt{(R^2 + X_L^2)} = 100$$
$$\sqrt{(100 + X_L^2)} = 100$$
$$100 + X_L^2 = 10\,000$$
$$X_L^2 = 10\,000 - 100$$
$$= 9900$$
$$X_L = 99.5 \, \Omega$$
$$2\pi f L = 99.5$$

$$f = \frac{99.5}{2\pi \times 20 \times 10^{-3}} \text{ Hz}$$
$$= 792 \text{ Hz}$$

$$\phi = \tan^{-1} \frac{99.5}{10} = 84.3° \text{ current lagging.}$$

2. A series circuit consists of a coil of inductance 2 H and resistance 100 Ω and a capacitor, C. Find the two values of C which will make the current 1.5 A when the circuit is connected to a 240 V, 50 Hz supply. What is the phase angle in each case?

Solution

$$Z = \frac{240}{1.5} = 160 \, \Omega$$
$$Z^2 = R^2 + X^2$$
$$X^2 = 160^2 - 100^2$$
$$= 15\,600$$
$$X = 125 \, \Omega$$
$$X_L = 2\pi f L$$
$$= 2\pi \times 50 \times 2$$
$$= 628 \, \Omega$$

$X_C = 503\,\Omega$ or $753\,\Omega$

$$C = \frac{1}{2\pi \times 50 \times 503}\,F \text{ or } \frac{1}{2\pi \times 50 \times 753}\,F$$

$= 6.3\,\mu F$ or $4.2\,\mu F$

$\phi = \tan^{-1}\dfrac{125}{100} = 51.3°.$

If $C = 6.3\,\mu F$ $X_L > X_C$ and the current lags by $51.3°$.
If $C = 4.2\,\mu F$ $X_L < X_C$ and the current leads by $51.3°$.

6B Test questions

1. A coil of inductance 5 mH and negligible resistance is connected to a 1 V, 2 MHz supply. Find the reactance of the coil, and the current flowing in magnitude and phase.

2. A perfect capacitor of 100 pF has a sinusoidal current flowing in it of 15 μA and a voltage across it of 0.1 V. Determine the frequency of the supply.

3. A coil of inductance 10 mH and resistance 100 Ω is connected to a 10 V (r.m.s.) 2 kHz supply. Find
   (a) the impedance of the coil,
   (b) the current flowing,
   (c) the phase angle,
   (d) the power dissipated.

4. A coil is connected to a 100 V, 50 Hz supply. It takes a current of 3 A and dissipates a power of 200 W. Find the resistance and inductance of the coil.

5. When 2 A d.c. passes through a coil the voltage across it is 20 V. If 2 A, 40 Hz passes through it the voltage across it is 80 V. The coil is now connected to a 240 V, 50 Hz supply. Find
   (a) the resistance of the coil,
   (b) the inductance of the coil,
   (c) the current flowing,
   (d) the phase angle,
   (e) the power is dissipated.

6. A 100 Ω resistor and a capacitor are in series across a 150 V, 200 Hz supply. If the current is 1 A find the capacitance of the capacitor.

7. A series circuit consists of a coil of resistance 350 Ω and inductance 100 mH and a 0.15 μF capacitor. If the supply is 2 V, 1 kHz find
   (a) the impedance of the circuit,
   (b) the current and its phase angle,
   (c) the power dissipated,
   (d) the voltage across the coil,
   (e) the voltage across the capacitor.

6.4 Series resonance

The impedance of the series circuit of Fig. 6.34 is given by eqn (6.36)

$$Z = \sqrt{[R^2 + (X_L - X_C)^2]}$$

At zero frequency (d.c.) X_C will be infinite and no current will flow. At an infinite frequency X_L will be infinite and, again, no current will flow. Somewhere between these two extremes the current must reach a maximum and this must occur when the impedance is a minimum. Eqn (6.36) shows that this happens when

$$X_L = X_C$$

and $\quad Z = \sqrt{R^2} = R.$

The current is then only limited by the resistance in the circuit, and if this is small, very large currents may flow. At this frequency current and voltage are in phase because, from eqn (6.35)

$$\phi = \tan^{-1} \frac{X_L - X_C}{R}$$

$$= \tan^{-1} \frac{0}{R}$$

$$= 0.$$

The circuit is said to be at *resonance*, and the frequency at which this occurs is called the *resonant frequency*, f_R. f_R can be found quite easily

$$X_L = X_C$$

$$2\pi f_R L = \frac{1}{2\pi f_R C}$$

$$f_R^2 = \frac{1}{4\pi^2 L C}$$

(6.40)
$$f_R = \frac{1}{2\pi \sqrt{(LC)}}.$$

In the series circuit, eqn (6.40) shows that the resonant frequency does not depend on the resistance in the circuit.

Resonance is a well-known phenomenon in mechanics. Consider a simple pendulum, that is a mass on the end of a string. If we hit the mass periodically various modes of vibration may occur. It is second nature to realize that if we hit it with a certain frequency, we shall be hitting it in phase with its own motion and the oscillations produced will be large. This is analogous to the electrical example and, in this case, the current will be a maximum. At frequencies below f_R the circuit will be capacitative because $X_C > X_L$. At frequency above f_R it will be inductive because $X_L > X_C$. This is shown in Fig. 6.37 where X_L and X_C are plotted against frequency. As can be seen, they must be equal at one frequency, f_R.

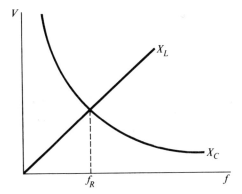

Fig. 6.37.

Fig. 6.38 shows graphs of Z and I against frequency.

As $Z = R$ at resonance the whole of the applied voltage appears across the resistor. Voltages, however, still exist across the inductor and capacitor, but they are of necessity equal and 180° out of phase and hence cancel (Fig. 6.39). The following example shows that they can be many times larger than the applied voltage.

Consider a series circuit with $R = 150\,\Omega$, $L = 10\,\text{mH}$, and $C = 50\,\text{pF}$. The resonant frequency is

$$f_R = \frac{1}{2\pi \sqrt{(LC)}}$$

$$= \frac{1}{2\pi \times \sqrt{(10 \times 10^{-3} \times 50 \times 10^{-12})}}\,\text{Hz}$$

$$= 225\,\text{kHz}.$$

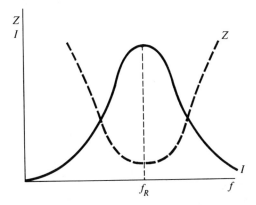

Fig. 6.38.

124 A.C. circuits I

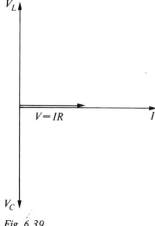

Fig. 6.39.

If a voltage of 1 V at this frequency is applied, the current will be

$$I = \frac{1}{150} \text{A} = 6.67 \,\text{mA}.$$

The inductive reactance at this frequency is

$$X_L = 2\pi \times 225 \times 10^3 \times 10 \times 10^{-3}$$
$$= 14\,137 \,\Omega.$$

Obviously X_C will be the same.
The voltage across the inductor is

$$V_L = I X_L$$
$$= 6.67 \times 10^{-3} \times 14\,137$$
$$= 94.3 \,\text{V}.$$

There will be an equal voltage, 180° out of phase, across the capacitor. The voltage across the coil (and the capacitor) is thus 94.3 times as large as the applied voltage, a quantity known as the magnification, or Q factor, of the circuit. This will be dealt with in greater detail in the next chapter.

6C Test questions

1. In a series L–R–C circuit, at resonance,
   (a) $X_L = R$,
   (b) $X_C = R$,
   (c) $X_L = X_C$,
   (d) the voltages across L and C are in phase.

2. In a series L–R–C circuit, at resonance,
   (a) the current is a maximum,
   (b) the current is a minimum,
   (c) the current is zero,
   (d) the current does not depend on R.

3. The resonant frequency of a series circuit is

(a) $\dfrac{1}{2\pi\, L\, C}$,

(b) $\dfrac{2\pi}{L\, C}$,

(c) $\dfrac{1}{\sqrt{(2\pi\, L\, C)}}$,

(d) $\dfrac{1}{2\pi\, \sqrt{(L\, C)}}$.

4. In a series L–R–C circuit it is found that the voltage across L is greater than the supply voltage. The circuit must be at resonance.
 (a) true,
 (b) false.

5. A coil and capacitor are in series across a 9 V r.m.s. sinusoidal supply. If the coil is 0.04 H, 450 Ω and the capacitor is 1 μF find
 (a) the resonant frequency,
 (b) the current at resonance,
 (c) the power at resonance,
 (d) the voltage across the capacitor at resonance.

7 A.C. circuits II

7.1 Power in a.c. circuits

We have seen in the previous chapter that there is no *average* power dissipated in a pure inductor or capacitor. In a resistor the power is given by

(7.1) $$P = I_{\text{r.m.s.}}^2 \times R$$

which follows from the definition of the r.m.s. value. Thus in a circuit containing R, L, and C in series, the power is $I^2 R$, I being the r.m.s. value. As we have seen

$$P = I^2 R$$
$$= \frac{V}{Z} I R$$
(7.2) $$= VI \cos \phi$$

because $R/Z = \cos \phi$ from the impedance triangle Fig. 7.1, which is similar to that shown in Fig. 6.33. It has been assumed in this diagram that the circuit is inductive and $X = X_L - X_C$.

The product of r.m.s. voltage and current is called the *volt-amperes* or *VA* of the circuit. It is larger than the power, unless the circuit is resistive when it is equal to it. The ratio of the actual power to the *VA* is called the *power factor* of the circuit. It is unity for a resistive circuit and zero for a reactive one.

(7.3) $$\text{power factor} = \frac{\text{power (watts)}}{VA}.$$

For the case of *sinusoidal quantities* the power factor is equal to $\cos \phi$ as can be seen from eqn (7.2).

As we saw in Chapter 6, a phasor can be resolved into two components at right angles. Consider the phasors of voltage and current (Fig. 7.2). The

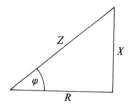

Fig. 7.1.

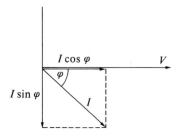

Fig. 7.2.

current is shown lagging the voltage by an angle ϕ. The components of the current are

$I \cos \phi$ in phase with V

$I \sin \phi$ 90° behind V.

The power is given by the product of the voltage and the component of I in phase with V. $I \cos \phi$ is the power producing or *active component* of the current. The other component, $I \sin \phi$, produces no power and is called the wattless or *reactive component*. The current, and its components, are shown in Fig. 7.3. They form a triangle similar to the impedance triangle. Fig. 7.4 shows a similar triangle with the sides multiplied by V. The sides are

$VI \cos \phi = $ power

$VI = VA$.

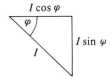

Fig. 7.3.

Fig. 7.4.

The third side, $VI \sin\phi$, is called the reactive VA or VAr of the circuit, and the triangle is called the power triangle. It should be noted that

$$(VAr)^2 + \text{power}^2 = V^2I^2 \sin^2\phi + V^2I^2 \cos^2\phi$$
$$= V^2I^2(\sin^2\phi + \cos^2\phi)$$
$$= V^2I^2$$
$$= (VA)^2.$$

This result can also be seen from the power triangle of Fig. 7.4.

Examples 7.1

1. A coil has a resistance of $12\,\Omega$ and an inductance of $0.05\,H$. It is connected to a $240\,V$, $50\,Hz$ supply. Find (a) the current, (b) the phase angle between voltage and current, (c) the power factor, (d) the VA, (e) the VAr, and (f) the power.

Solution

$$X_L = 2\pi fL$$
$$= 2 \times \pi \times 50 \times 0.05$$
$$= 15.7\,\Omega.$$

$$Z = \sqrt{(R^2 + X_L^2)}$$
$$= \sqrt{(12^2 + 15.7^2)}$$
$$= 19.8\,\Omega.$$

(a) $I = V/Z$

$$= \frac{240}{19.8}$$

$$= 12.1\,A.$$

(b) $\phi = \tan^{-1} \dfrac{15.7}{12}$

$= 52.6°$ current lagging.

(c) Power factor $= \cos\phi$
$= 0.607.$

(d) $VA = 240 \times 12.1 = 2904\,VA.$

(e) $VAr = VA \sin\phi$
$= 2904 \sin 52.6°$
$= 2307\,VAr.$

(f) $P = I^2 R$
$= 12.1^2 \times 12 = 1757\,W.$

2. A motor is developing 20 b.h.p. with an efficiency of 80 per cent. If the motor operates from a $240\,V$, $50\,Hz$ supply with a power factor of 0.6 lagging, find

(a) the input VA,
(b) the active current,
(c) the reactive current,
(d) the input VAr.

Solution

(a) 20 b.h.p. = 20 × 746 = 14 920 W.

This is the output power.

$$\text{Input power} = \frac{14\,920}{0.8} = 18\,650\,\text{W}.$$

$$\text{Input } VA = \frac{18\,650}{0.6} = 31\,083\, VA.$$

(b) $\quad$ input current $= \dfrac{31\,083}{240} = 129.5$ A.

active component = 129.5 × 0.6
$\qquad\qquad\qquad\;\;= 77.7$ A.

(c) $\qquad\qquad\qquad \phi = \cos^{-1} 0.6 = 53.1°$

$\qquad\qquad\qquad \sin\phi = 0.8,$

reactive current = 129.5 × 0.8
$\qquad\qquad\qquad = 103.6$ A.

(Note that

$\qquad 103.6^2 + 77.7^2 = 129.5^2$).

(d) Input $VAr = VA \sin\phi$
$\qquad\qquad\quad = 31\,083 \times 0.8 = 24\,866\, VAr.$

3. An a.c. generator develops 500 V and has a maximum current output of 250 A. Find the power it can supply when feeding a load with a power factor of (a) zero, (b) unity, and (c) 0.6.

Solution

(a) VA of generator = 500 × 250
$\qquad\qquad\qquad\quad\;\;= 125\,000$ VA
$\qquad\qquad\qquad\quad\;\;= 125$ kVA.

$\quad$ Power factor zero

$\qquad$ power = 125 × 0 = 0.

(b) Power = 125 × 1 = 125 kW.
(c) Power = 125 × 0.6 = 75 kW.

130   A.C. circuits II

4. A series circuit has $R = 10\,\Omega$, $L = 30\,\text{mH}$, and $C = 150\,\mu\text{F}$. Find, for an applied voltage of 200 V, 50 Hz
   (a) the current,
   (b) the active current,
   (c) the reactive current,
   (d) the power factor,
   (e) the power, VA, and $VA\text{r}$.

Solution

(a) $X_L = 2 \times \pi \times 50 \times 30 \times 10^{-3}$
$= 9.4\,\Omega$.

$X_C = \dfrac{1}{2 \times \pi \times 50 \times 150 \times 10^{-6}}$

$= 21.2\,\Omega$.

$X = 21.2 - 9.4 = 11.8\,\Omega$ capacitive.

$Z = \sqrt{(10^2 + 11.8^2)}$
$= 15.5\,\Omega$.

$I = \dfrac{200}{15.5} = 12.9\,\text{A}$ leading by

$\phi = \tan^{-1} \dfrac{11.8}{10} = 50°$.

(b) Active current $= 12.9 \cos 50 = 8.29\,\text{A}$.
(c) Reactive current $= 12.9 \sin 50 = 9.88\,\text{A}$.
(d) Power factor $= \cos 50 = 0.64$.
(e) $P = 200 \times 12.9 \times 0.64 = 1651\,\text{W}$.
$VA = 200 \times 12.9 = 2580\,VA$.
$VA\text{r} = 2580 \sin 50 = 1976\,VA\text{r}$.

7A Test questions

1. Power factor is defined as
   (a) $\dfrac{\text{power}}{VA}$,
   (b) $\cos\phi$,
   (c) $\sin\phi$,
   (d) $\dfrac{VA}{\text{power}}$.

2. The phase angle between a sinusoidal voltage and current is 30°. The power factor is
   (a) unity,
   (b) depends on whether the current leads or lags,

(c) 0.866,
(d) 0.5.

3. A series circuit has a *VA* of 1000 VA and a *VA*r of 500 VAr. Determine the power being developed.

4. If the resistance in the circuit of question 3 is 50 Ω, find the current flowing.

5. A coil of resistance 10 Ω and inductance 0.1 H is connected to a 240 V, 50 Hz supply. Find
  (a) the current in magnitude and phase,
  (b) the power factor,
  (c) the *VA*,,
  (d) the *VA*r,
  (e) the power.

6. A 15 Ω resistor and a capacitor in series are connected to a 50 Hz supply. The voltage across the components are 30 V and 40 V respectively. Find
  (a) the capacitance,
  (b) the supply voltage,
  (c) the power developed,
  (d) the power factor.

7. A coil takes 0.8 kW and 1.5 kVAr when connected to a 240 V, 50 Hz supply. Find the resistance and inductance of the coil.

8. The coil of question 7 is connected in series with a 200 μF capacitor to the same supply. Find
  (a) the current in magnitude and phase,
  (b) the power,
  (c) the *VA*r,
  (d) the *VA*.

7.2 Q factor

We saw in the last chapter that a series R-L-C circuit has a maximum current at a frequency f_R such that

(7.4) $$f_R = \frac{1}{2\pi \sqrt{(LC)}}.$$

At this frequency $X_L = X_C$ and the voltage and current are in phase. The latter is, in fact, the definition of resonance in an electrical circuit. (Current is not a maximum at resonance in all circuits!) The current at resonance in a series circuit is given by

(7.5) $$I = \frac{V}{R}.$$

The voltages across L and C will, at resonance, be equal and 180° out of phase. There is zero resultant voltage across L and C combined, all the applied voltage appearing across R. (We are assuming here that any resistance possessed by the coil is included in R and that L represents the purely inductive part of the coil.)

The voltage across L at resonance is

$$V_L = 2\pi f_R LI$$
$$\quad = \omega_R LI. \tag{7.6}$$

This voltage, equal in magnitude to the voltage across the capacitor, may well be larger than the applied voltage, perhaps many times larger. The ratio of the voltage across L (or C) to the applied voltage is called the Q factor (quality factor) of the circuit. Q is sometimes called the voltage magnification. Clearly

$$Q = \frac{\omega_R LI}{V} = \frac{\omega_R L}{R}. \tag{7.7}$$

The voltage across C is

$$V_C = \frac{I}{\omega_R C}$$

hence Q is also given by

$$Q = \frac{I}{\omega_R CV} = \frac{1}{\omega_R CR} \tag{7.8}$$

and as

$$\omega_R = \frac{1}{\sqrt{(LC)}}$$

Q may also be written

$$Q = \frac{1}{R} \sqrt{(L/C)} \tag{7.9}$$

from eqns (7.7) or (7.8).

The real meaning of Q can best be illustrated using a numerical example. Consider a series circuit with $R = 10\,\Omega$, $L = 1$ H, and a value of C to produce resonance at 50 Hz.

$$Q = \frac{\omega_R L}{R}$$

$$= \frac{2\pi\, 50 \times 1}{10}$$

$$= 31.4.$$

If this circuit is connected to a 100 V supply, the current at resonance will be 10 A. If the supply frequency is varied, the current at different frequencies will be as shown in Fig. 7.5. These values have been found by using a value of C of $10.13\,\mu$F, the capacitance required to produce

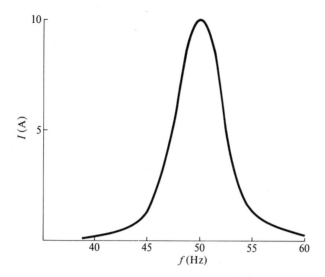

Fig. 7.5.

resonance at 50 Hz. This curve represents a circuit with a Q of just over 30. Fig. 7.6 shows the same curve plotted with two others, representing Qs of 2 and 50 respectively. (They correspond to inductance values of 0.064 H and 1.59 H respectively.) It can be seen that the higher the value of Q, the greater is the ability of the circuit to select one frequency, or to be more exact a narrow band of frequencies.

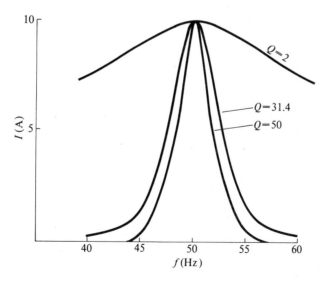

Fig. 7.6.

Examples 7.2

1. A series circuit consists of a capacitor of 50 pF and a coil of inductance 100 μH and resistance 25 Ω. Find the Q of the circuit and the resonant frequency.

Solution

$$Q = \frac{1}{R}\sqrt{(L/C)}$$

$$= \frac{1}{25}\sqrt{(100 \times 10^{-6}/50 \times 10^{-12})}$$

$$= 56.6.$$

$$f_R = \frac{1}{2\pi\sqrt{(LC)}}$$

$$= \frac{1}{2\pi\sqrt{(100 \times 10^{-6} \times 50 \times 10^{-12})}} \text{ Hz}$$

$$= 2.25 \text{ MHz},$$

or
$$Q = \frac{2\pi f_R L}{R}$$

$$f_R = \frac{QR}{2\pi L}$$

$$= \frac{56.6 \times 25}{2\pi\, 100 \times 10^{-6}} \text{ Hz}$$

$$= 2.25 \text{ MHz}.$$

2. A series circuit consists of a coil of inductance 0.3 H and resistance 25 Ω and a capacitor. An applied voltage of 240 V gives a maximum current when the frequency is 50 Hz. Find
  (a) the value of C,
  (b) the current at 50 Hz,
  (c) the voltage across the capacitor at 50 Hz,
  (d) the voltage across the coil at 50 Hz,
  (e) the power dissipated at 50 Hz,
  (f) the Q factor of the circuit.

Solution

(a) $f_R = \dfrac{1}{2\pi\sqrt{(LC)}}$

$$C = \frac{1}{4\pi^2 f_R^2 L}$$

$$= \frac{1}{4\pi^2 \; 50^2 \times 0.3} \; F$$

$$= 33.8 \, \mu F.$$

(b) $I = \dfrac{240}{25} = 9.6 \, A.$

(c) $V_C = \dfrac{I}{\omega_R C}$

$$= \frac{9.6}{2\pi \; 50 \times 33.8 \times 10^{-6}}$$

$$= 904.1 \, V.$$

(d) At 50 Hz the coil has an impedance of

$$Z = \sqrt{(25^2 + 4\pi^2 \; 50^2 \times 0.3^2)}$$
$$= 97.5 \, \Omega.$$

$V_{coil} = 9.6 \times 97.5$
$\phantom{V_{coil}} = 936 \, V.$

(e) $P = I^2 R$
$ = 9.6^2 \times 25$
$ = 2304 \, W.$

(f) $Q = \dfrac{V_C}{240} = \dfrac{904.1}{240}$

$ = 3.77.$

3. A coil has a resistance of 50 Ω and an inductance of 250 μH. Find the value of the capacitance in series with it which produces a Q factor of 75. What is the resonant frequency of this circuit?

Solution

$$Q = \frac{1}{R} \sqrt{(L/C)}$$

$$C = \frac{L}{Q^2 \times R^2}$$

$$= \frac{250 \times 10^{-6}}{75^2 \times 50^2} \; F$$

$$= 17.8 \, pF.$$

136 A.C. circuits II

$$f_R = \frac{1}{2\pi\sqrt{(LC)}}$$

$$= \frac{1}{2\pi\sqrt{(250 \times 10^{-6} \times 17.8 \times 10^{-12})}} \text{ Hz}$$

$$= 2.39 \text{ MHz}.$$

7B Test questions

1. A series circuit has $L = 4\,\mu\text{H}$, $C = 1\,\mu\text{F}$, and $R = 0.1\,\Omega$. The Q factor is
   (a) 20,
   (b) 5,
   (c) 40,
   (d) 1.

2. A circuit consists of a coil of resistance $50\,\Omega$ and inductance $75\,\mu\text{H}$ in series with a $100\,\text{pF}$ capacitor. Find the resonant frequency and Q factor.

3. An inductor of negligible resistance is in series with a $200\,\text{pF}$ capacitor and a $100\,\Omega$ resistor. If the resonant frequency is $150\,\text{kHz}$ find the value of L and the Q factor of the circuit.

4. A coil of inductance $100\,\mu\text{H}$ and resistance $15\,\Omega$ is in series with a capacitor. If the Q factor of the circuit is 60 find the value of the capacitance and the resonant frequency of the circuit.

5. A coil of resistance $16\,\Omega$ and inductance $0.095\,\text{H}$ is in series with a $200\,\mu\text{F}$ capacitor. Find the resonant frequency and Q factor.

6. The Q factor of the circuit of question 5 could be increased by
   (a) increasing the resistance,
   (b) increasing the capacitance,
   (c) increasing the capacitance and inductance in the same ratios,
   (d) increasing L and decreasing C.

7.3 Parallel circuits

C and R in parallel

It is quite possible to obtain resistors and capacitors that are to all intents and purposes perfect, so that this combination represents a practical case. The circuit is given in Fig. 7.7 and the phasor diagram in Fig. 7.8. The impedance of the circuit is, of course, V/I. From the phasor diagram

$$I^2 = I_R^2 + I_C^2$$

$$= \frac{V^2}{R^2} + \omega^2 C^2 V^2$$

$$= V^2\left(\frac{1}{R^2} + \omega^2 C^2\right)$$

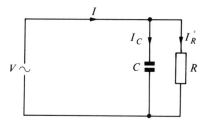

Fig. 7.7.

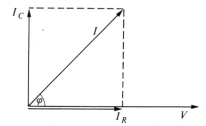

Fig. 7.8.

$$Z = \frac{V}{I}$$

$$= \sqrt{\left(\frac{1}{\frac{1}{R^2} + \omega^2 C^2}\right)}$$

$$= \sqrt{\left(\frac{R^2}{1 + \omega^2 C^2 R^2}\right)}$$

(7.10)
$$= \frac{R}{\sqrt{(1 + \omega^2 C^2 R^2)}}.$$

The phase angle $\phi = \tan^{-1} \dfrac{I_C}{I_R}$

$$= \tan^{-1} \frac{\omega CV}{V/R}$$

(7.11)
$$= \tan^{-1} \omega CR \text{ (current leading)}.$$

Consider values of C and R (Fig. 7.7) of $10\,\mu\text{F}$ and $300\,\Omega$ respectively and that the supply is 240 V, 50 Hz.

$$X_C = \frac{1}{2\pi\, 50 \times 10 \times 10^{-6}}$$

$$= 318\,\Omega.$$

$$I_C = \frac{240}{318}$$

$$= 0.755 \text{ A leading } V \text{ by } 90°.$$

$$I_R = \frac{240}{300}$$

$$= 0.8 \text{ A in phase with } V.$$

Obviously, from Fig. 7.8

$$I = \sqrt{(0.8^2 + 0.755^2)}$$
$$= 1.1 \text{ A}$$

and

$$\phi = \tan^{-1}\frac{0.755}{0.8}$$

$$= 43.3°.$$

Hence the total current is 1.1 A leading V by 43.3°.
The impedance of the circuit is

$$Z = \frac{V}{I}$$

$$= \frac{240}{1.1}$$

$$= 218.2 \text{ }\Omega.$$

The total current could have been found by calculating Z from eqn (7.10).

$$Z = \frac{R}{\sqrt{(1 + \omega^2 C^2 R^2)}}$$

$$= \frac{300}{\sqrt{(1 + 4\pi^2 \, 50^2 \times 10^2 \times 10^{-12} \times 300^2)}}$$

$$= 218.2 \text{ }\Omega$$

giving a current of

$$I = \frac{240}{218.2}$$

$$= 1.1 \text{ A},$$

leading V by

$$\phi = \tan^{-1}(2\pi \, 50 \times 10 \times 10^{-6} \times 300)$$
$$= \tan^{-1} 0.942$$
$$= 43.3°.$$

The power dissipated in the circuit is

$$P = I_R^2 R$$
$$= 0.8^2 \times 300$$
$$= 192 \text{ W}.$$

Note that as power is only dissipated in the resistor it must equal the current in the resistor squared multiplied by R. It could also have been calculated as the product of the total current, the voltage, and the cosine of the phase angle thus

$$P = 240 \times 1.1 \times \cos 43.3$$
$$= 192 \text{ W}.$$

The total current in this circuit can never be in phase with the applied voltage so that the circuit has no resonant frequency.

C *and* L *in parallel*

Fig. 7.9 shows a perfect inductor, L, in parallel with a perfect capacitor, C, connected to an alternating supply of voltage V. If the supply frequency is f, the reactances are

$$X_L = 2\pi f L = \omega L$$

$$X_C = \frac{1}{2\pi f C} = \frac{1}{\omega C}.$$

The current in L is given by

$$I_L = \frac{V}{X_L} = \frac{V}{\omega L} \text{ lagging } V \text{ by } 90°.$$

and in C by

$$I_C = \frac{V}{X_C} = \omega C V \text{ leading } V \text{ by } 90°.$$

The total current is the phasor sum of these, and, as they are 180° apart, is given by

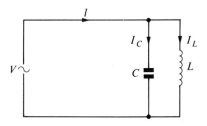

Fig. 7.9.

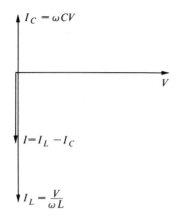

Fig. 7.10.

$$I = \frac{V}{\omega L} - \omega C V$$

(7.10)
$$= V\left(\frac{1}{\omega L} - \omega C\right).$$

The phasor diagram is shown in Fig. 7.10.

As an example consider a coil of inductance 2 H (no resistance) in parallel with a 3 μF capacitor connected to a 100 V, 50 Hz supply.

$$X_L = 2\pi\, 50 \times 2 = 628.3\, \Omega$$

and $$I_L = \frac{100}{628.3} = 0.159\text{ A lagging } V \text{ by } 90°.$$

$$X_C = \frac{10^6}{2\pi\, 50 \times 3} = 1061\, \Omega$$

$$I_C = \frac{100}{1061} = 0.094\text{ A leading } V \text{ by } 90°.$$

Hence $I = 0.159 - 0.094$
$\quad\quad\; = 0.065$ A lagging V by $90°$.

Note that each of the branch currents is larger than the total current. The impedance of the parallel circuit is

$$Z = \frac{V}{I} = \frac{100}{0.065} = 1538\, \Omega.$$

The frequency of the supply is now altered such that $X_L = X_C$. This occurs at a frequency given by eqn (7.4).

$$f_R = \frac{1}{2\pi\, \sqrt{(2 \times 3 \times 10^{-6})}}$$

$$= 65\text{ Hz}.$$

$$X_L = X_C = 2\pi\,65 \times 2$$
$$= 816.8\,\Omega$$

and $\quad I_L = I_C = \dfrac{100}{816.8} = 0.122\,\text{A}.$

The total current is

$$I = I_L - I_C = 0.$$

Power cannot be dissipated in any of these examples as there is no resistance in the circuit.

The last example, when $X_L = X_C$, is, in fact, resonance, and in this case occurs at the same frequency as it would if the two components were in series. The total current is, however, zero at resonance, in other words the impedance is infinite. It should be noted that no circuit can contain zero resistance so that this example is somewhat theoretical.

Capacitor and coil in parallel

Here we are considering the more practical case of a coil with resistance in parallel with a perfect capacitor. As explained earlier, capacitors can be obtained which are virtually perfect. Fig. 7.11 shows such a circuit connected to an a.c. supply, V.

The current in the capacitor, I_C, is, as before, given by

(7.11) $\qquad I_C = \omega C V$ leading V by $90°$.

The coil has an impedance Z_1 given by

$$Z_1 = \sqrt{(R^2 + \omega^2 L^2)}$$

so that

(7.12) $\qquad I_1 = \dfrac{V}{Z_1} = \dfrac{V}{\sqrt{(R^2 + \omega^2 L^2)}}$

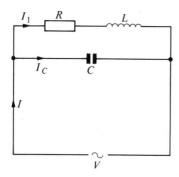

Fig. 7.11.

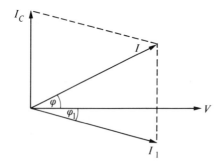

Fig. 7.12.

lagging V by ϕ_1 where

(7.13) $$\phi_1 = \tan^{-1} \frac{\omega L}{R}.$$

These currents are shown on a phasor diagram in Fig. 7.12. The diagram also shows the total, or supply, current I as the phasor sum of I_C and I_1. It is shown leading V by an angle ϕ_1, although it could, of course, be lagging.

It is possible to produce a formula for the impedance of the circuit of Fig. 7.11, but it is a somewhat complicated expression. It is simpler to calculate the currents I_C and I_1 and add them either by drawing a scale phasor diagram, or by resolving the currents into components, as will be seen in Examples 7.3.

As with the case of L and C in parallel, it is possible for the current in each branch to be greater than the supply current. Fig. 7.13 shows a phasor diagram of such a situation. In this case the supply current is lagging the applied voltage. It is also possible for the supply current to be in phase with the voltage. (It is almost so in Fig. 7.13.) This would be resonance, and will be dealt with in Section 7.4.

Examples 7.3

1. A capacitor of 100 pF is in parallel with a 500 Ω resistor. The parallel combination is connected to a 0.5 V, 5 MHz supply. Determine the total current taken from the supply and its phase angle relative to the applied voltage. What power is taken from the supply?

Solution

$$X_C = \frac{1}{2\pi \times 5 \times 10^6 \times 100 \times 10^{-12}}$$
$$= 318.3 \ \Omega$$

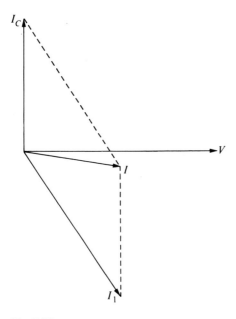

Fig. 7.13.

$$I_C = \frac{0.5}{318.3} \text{ A}$$

$$= 1.57 \text{ mA } 90° \text{ ahead of } V.$$

$$I_R = \frac{0.5}{500} \text{ A}$$

$$= 1 \text{ mA in phase with } V \text{ (Fig. 7.14)}.$$

$$\text{Total current } I = \sqrt{(1^2 + 1.57^2)}$$
$$= 1.86 \text{ mA}$$

leading V by $\phi = \tan^{-1} \frac{1.57}{1}$

$$= 57.5°.$$

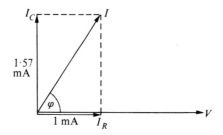

Fig. 7.14.

$$\text{Power} = I_R{}^2 R$$
$$= (1.10^{-3})^2 \times 500 \text{ W}$$
$$= 500 \ \mu\text{W},$$

or
$$P = VI \cos \phi$$
$$= 0.5 \times 1.86 \times 10^{-3} \times \cos 57.5 \text{ W}$$
$$= 500 \ \mu\text{W}.$$

2. A circuit consisting of a capacitor and resistor in parallel takes a total current of 3.5 A from a 150 V, 1 kHz supply. If the power taken from the supply is 450 W, determine the value of the capacitor.

Solution

$$P = \frac{V^2}{R}$$

$$\therefore \quad R = \frac{V^2}{P}$$

$$= \frac{150^2}{450}$$

$$= 50 \ \Omega.$$

$$I_R = \frac{150}{50}$$

$$= 3 \text{ A}.$$

$$I_C = \sqrt{(I^2 - I_R{}^2)}$$
$$= \sqrt{(3.5^2 - 3^2)}$$
$$= 1.8 \text{ A}.$$

$$X_C = \frac{150}{1.8}$$

$$= 83.3 \ \Omega.$$

$$C = \frac{1}{2\pi f X_C}$$

$$= \frac{1}{2\pi \times 10^3 \times 83.3} \text{ F}$$

$$= 1.9 \ \mu\text{F}.$$

Alternate solution

$$Z = \frac{R}{\sqrt{(1 + \omega^2 C^2 R^2)}} \qquad \text{(eqn. 7.10)}$$

Parallel circuits 145

$$Z^2(1 + \omega^2 C^2 R^2) = R^2$$
$$\omega^2 C^2 R^2 Z^2 = R^2 - Z^2$$
$$C^2 = \frac{R^2 - Z^2}{\omega^2 R^2 Z^2}$$
$$C = \sqrt{\left(\frac{R^2 - Z^2}{\omega^2 R^2 Z^2}\right)}$$
$$= \sqrt{\left(\frac{50^2 - 42.9^2}{4\pi^2 (1000)^2 \times 50^2 \times 42.9^2}\right)} F$$
$$= 1.9 \, \mu F.$$

3. A perfect inductor of $10 \, \mu H$ is in parallel with a capacitor. If the circuit resonates at 15 MHz, find the value of the capacitor, the current taken by each component, and the average power dissipated. The supply voltage is 10 mV.

Solution

$$X_L = 2\pi \times 15 \times 10^6 \times 10 \times 10^{-6}$$
$$= 942.5 \, \Omega$$

$\therefore \quad X_C = 942.5 \, \Omega.$

$$C = \frac{1}{2\pi \times 15 \times 10^6 \times 942.5} F$$
$$= 11.3 \, pF.$$

$$I_L = \frac{10 \times 10^{-3}}{942.5} A$$
$$= 10.6 \, \mu A \quad 90° \text{ behind } V.$$

Hence $I_C = 10.6 \, \mu A \quad 90°$ ahead of V.

The average power is zero.

4. Find the total current in the circuit of question 3 if the supply frequency is changed to 10 kHz.

Solution

$$X_L = 2\pi \times 10 \times 10^3 \times 10 \times 10^{-6}$$
$$= 0.63 \, \Omega.$$

$$I_L = \frac{10 \times 10^{-3}}{0.63} A$$
$$= 15.9 \, mA.$$

146 A.C. circuits II

$$X_C = \frac{1}{2\pi \times 10 \times 10^3 \times 11.3 \times 10^{-12}} \,\Omega$$
$$= 1.4 \,\text{M}\Omega.$$

$$I_C = \frac{10 \times 10^{-3}}{1.4 \times 10^6} \,\text{A}$$
$$= 7.14 \times 10^{-9} \,\text{A} \text{ or } 7.14 \,\text{nA}.$$

I_C is negligible compared with I_L, hence the total current is 15.9 mA, 90° behind the voltage.

5. A coil of inductance 2 mH and resistance 3 Ω is in parallel with a 40 μF capacitor. They are connected to a 10 V, 400 Hz supply. Find the current in each branch and the total current. What is the impedance of the circuit at 400 Hz?

Solution

$$X_L = 2\pi \times 400 \times 2 \times 10^{-3}$$
$$= 5 \,\Omega.$$

$$Z_1 = \sqrt{(3^2 + 5^2)}$$
$$= 5.83 \,\Omega.$$

$$I_1 = \frac{10}{5.83}$$
$$= 1.72 \,\text{A}.$$

This lags the applied voltage by

$$\phi_1 = \tan^{-1} \tfrac{5}{3}$$
$$= 59°.$$

$$X_C = \frac{1}{2\pi \times 400 \times 40 \times 10^{-6}}$$
$$= 9.9 \,\Omega.$$

$$I_C = \frac{10}{9.9}$$
$$= 1.01 \,\text{A leading by } 90°.$$

Hence the phasor diagram is as Fig. 7.15.
I_1 may be resolved into

$$1.72 \cos 59° = 0.89 \,\text{A horizontal}$$
$$1.72 \sin 59° = 1.47 \,\text{A downwards}.$$

$$\text{Total vertical component} = 1.47 - 1.01$$
$$= 0.46 \,\text{A downwards (Fig. 7.16)}.$$

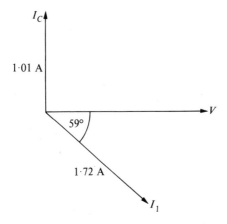

Fig. 7.15.

$$I = \sqrt{(0.89^2 + 0.46^2)}$$
$$= 1 \text{ A}.$$

$$\phi = \tan^{-1} \frac{0.46}{0.89}$$
$$= 27.3°.$$

Total current is 1 A lagging the voltage by 27.3°.

$$\text{Impedance} = \frac{V}{I}$$
$$= \frac{10}{1}$$
$$= 10 \, \Omega.$$

6. A coil of inductance $10 \, \mu\text{H}$ and resistance $100 \, \Omega$ is in parallel with a capacitor, C. When connected to a 1 V, 2 MHz supply the total current is found to be 5 mA. Find the two possible values of C.

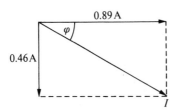

Fig. 7.16.

148 A.C. circuits II

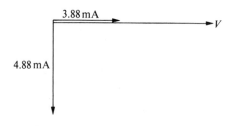

Fig. 7.17.

Solution

$$X_L = 2\pi \times 2 \times 10^6 \times 10 \times 10^{-6}$$
$$= 125.7 \, \Omega.$$

$$Z_1 = \sqrt{(100^2 + 125.7^2)}$$
$$= 160.6 \, \Omega.$$

$$I_1 = \frac{1}{160.6} \, A$$
$$= 6.23 \, mA.$$

$$\phi_1 = \tan^{-1} \frac{125.7}{100}$$
$$= 51.5° \text{ lagging}.$$

I_1 has components (Fig. 7.17).

$$I_{1H} = 6.23 \cos 51.5 = 3.88 \, mA$$
$$I_{1V} = 6.23 \cos 51.5 = 4.88 \, mA.$$

The total current (5 mA) consists of this coil current added to the capacitor current which leads V by 90°. The total vertical component (which adds to the horizontal component of 3.88 mA to give 5 mA) must be given by

$$I_V = \sqrt{(5^2 - 3.88^2)}$$
$$= 3.15 \, mA.$$

This might be leading or lagging V by 90°.
  Case (a)—leading (Fig. 7.18)

$$I_C = 4.88 + 3.15$$
$$= 8.03 \, mA.$$

$$X_C = \frac{1}{8.03 \times 10^{-3}}$$
$$= 124.5 \, \Omega.$$

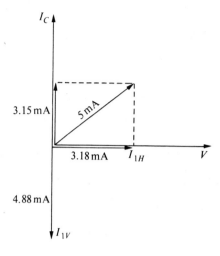

Fig. 7.18.

$$C = \frac{1}{2\pi \times 2 \times 10^6 \times 124.5} \text{ F}$$
$$= 639 \text{ pF}.$$

Case (b)—lagging (Fig. 7.19).

$$I_C = 4.88 - 3.15$$
$$= 1.73 \text{ mA}.$$

$$X_C = \frac{1}{1.73 \times 10^{-3}}$$
$$= 578 \text{ }\Omega.$$

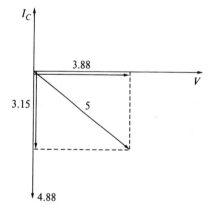

Fig. 7.19.

150 A.C. circuits II

$$C = \frac{1}{2\pi \times 2 \times 10^6 \times 578} \text{ F}$$

$$= 137.7 \text{ pF}.$$

7C Test questions

1. A perfect inductor of inductance $100\,\mu\text{H}$ is in parallel with a 15 pF capacitor. Determine
   (a) the current in each component and the total current when connected to a 2 V, 5 MHz supply,
   (b) the phase angle between the voltage and total current in (a),
   (c) the resonant frequency of the circuit,
   (d) the current in each component if the supply is 2 V at the resonant frequency,
   (e) the Q factor.

2. A $1000\,\Omega$ resistor is in parallel with a capacitor, C. The total current taken from a 50 V, 2000 Hz supply is 120 mA. Find the value of C, the power developed, and the phase angle between the applied voltage and the total current.

3. A perfect 2 H inductor is connected to a 240 V, 50 Hz supply. Find the capacitance in parallel with the inductor which will reduce the total current to 0.25 A. (Two answers.)

4. A coil of inductance $100\,\mu\text{H}$ and resistance $2000\,\Omega$ is in parallel with a 15 pF capacitor. If the circuit is connected to a 2.5 V, 5 MHz supply, find
   (a) the current in each component,
   (b) the total current in magnitude and phase,
   (c) the impedance of the circuit at 5 MHz,
   (d) the power developed.

7.4 Parallel resonance

The parallel circuit of Fig. 7.11 consists of a real coil (R and L) in parallel with a perfect capacitor, C. We have seen in the previous section that the total current may lead or lag the applied voltage. It may also be in phase with it, which is the resonant condition. Fig. 7.20 is the phasor diagram of such a situation. The resonant frequency will be designated f_0 to distinguish it from the resonant frequency of R, L, and C in series which we have called f_R.

Now, as I_C has no horizontal component, I must equal $I_1 \cos\phi_1$. Also, as the total current I has no vertical component, I_C must equal, and be $180°$ out of phase with, the vertical component of I_1. Hence,

(7.14) $\quad I_C = I_1 \sin\phi_1.$

But $I_C = \omega_0 CV \quad$ (eqn (7.11))

and $I_1 = \dfrac{V}{Z_1} \quad$ (eqn (7.12)).

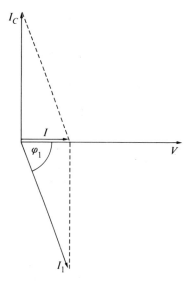

Fig. 7.20.

Also $\sin \phi_1 = \dfrac{\omega_0 L}{Z_1}$

from the impedance triangle of the coil.
Hence
$$\omega_0 C V = \dfrac{V}{Z_1} \times \dfrac{\omega_0 L}{Z_1}$$

(7.15) and $C = \dfrac{L}{Z_1^2}.$

This, then, is the condition for resonance
$$C Z_1^2 = L.$$
But $Z_1^2 = R^2 + \omega_0^2 L^2$
so that $C(R^2 + \omega_0^2 L^2) = L$

$$R^2 + \omega_0^2 L^2 = \dfrac{L}{C}$$

$$\omega_0^2 L^2 = \dfrac{L}{C} - R^2$$

$$\omega_0^2 = \dfrac{1}{LC} - \dfrac{R^2}{L^2}$$

$$\omega_0 = \sqrt{\left(\dfrac{1}{LC} - \dfrac{R^2}{L^2}\right)}$$

152 A.C. circuits II

(7.16) $\quad$ or $\quad f_0 = \dfrac{1}{2\pi} \sqrt{\left(\dfrac{1}{LC} - \dfrac{R^2}{L^2}\right)}.$

It should be observed that if $R = 0$ eqn (7.16) becomes

$$f_0 = \dfrac{1}{2\pi} \sqrt{\left(\dfrac{1}{LC}\right)}$$

the same as f_R. This was found to be the case in Section 7.3 when considering L and C in parallel.

The resonant frequency of the coil in parallel with the capacitor is thus seen to be lower than that of the same coil in series with the capacitor, although the difference is often very small. For example, a coil of inductance 0.1 mH in series with a 100 pF capacitor has a resonant frequency of

$$f_R = \dfrac{1}{2\pi \sqrt{(0.1 \times 10^{-3} \times 100 \times 10^{-12})}} \text{ Hz}$$

$$= 1.592 \text{ MHz}$$

If the coil has a resistance of, say, 500 Ω, the resonant frequency of the coil and capacitor in parallel is

$$f_0 = \dfrac{1}{2\pi} \sqrt{\left(\dfrac{1}{0.1 \times 10^{-3} \times 100 \times 10^{-12}} - \dfrac{500^2}{(0.1 \times 10^{-3})^2}\right)}$$

$$= \dfrac{1}{2\pi} \sqrt{(10^{14} - 2.5 \times 10^{13})} \text{ Hz}$$

$$= 1.378 \text{ MHz} \text{ (13 per cent lower than } f_R\text{).}$$

However, if the coil has a resistance of only, say, 10 Ω

$$\dfrac{R^2}{L^2} = \dfrac{10^2}{(0.1 \times 10^{-3})^2}$$

$$= 10^{10}$$

and $\quad f_0 = \dfrac{1}{2\pi} \sqrt{(10^{14} - 10^{10})} \text{ Hz}$

$$= 1.591 \text{ MHz} \text{ (only 0.06 per cent lower than } f_R\text{).}$$

The total current I (Fig. 7.20) may be small and, in fact, becomes zero at resonance if R is zero. The parallel circuit of Fig. 7.11 has a high impedance at resonance, called the *dynamic impedance* of the circuit, Z_D. (It is usually referred to as an impedance although, because V and I are in phase, it is resistive.)

(7.17) $\quad Z_D = \dfrac{V}{I}.$

We have seen that

$$I_C = I_1 \sin \phi_1. \text{ (eqn (7.14))}$$

Parallel resonance

But $I = I_1 \cos \phi_1$ from Fig. 7.20.

Hence, dividing

$$\frac{I_C}{I} = \frac{I_1 \sin \phi_1}{I_1 \cos \phi_1}$$

$$= \tan \phi_1$$

$$= \frac{\omega_0 L}{R} \text{ from the impedance triangle.}$$

Also

$$I_C = \omega_0 C V \quad (\text{eqn (7.11)})$$

$$\frac{\omega_0 C V}{I} = \frac{\omega_0 L}{R}$$

(7.18) and $Z_D = \dfrac{V}{I} = \dfrac{L}{CR}.$

Eqn (7.18) shows that the dynamic impedance of the circuit *increases* as *R decreases*, becoming infinite if $R = 0$ (see Section 7.3). This is simply because the lower the resistance, the less power is dissipated, hence the less the power which needs to be supplied from the supply. The currents flowing round the parallel circuit are caused by the capacitor discharging via the coil. When the capacitor voltage has fallen to zero, the energy which it had originally stored has been transferred to the coil, some being dissipated in the resistance, the rest stored in the inductance. This stored energy keeps the current flowing (Lenz's Law) and the capacitor recharges with the opposite polarity. This action continues, like a swinging pendulum, the supply only needing to replace the energy lost in the resistance. In the limit if $R = 0$, no energy is lost and none need be supplied by the supply. Hence the total (supply) current is zero and the dynamic impedance infinite.

In the above example

if $R = 500\,\Omega$ $\quad Z_D = \dfrac{0.1 \times 10^{-3}}{100 \times 10^{-12} \times 500}\,\Omega = 2\,\text{k}\Omega$

if $R = 10\,\Omega$ $\quad Z_D = \dfrac{0.1 \times 10^{-3}}{100 \times 10^{-12} \times 10}\,\Omega = 100\,\text{k}\Omega.$

As can be seen in Fig. 7.20, the currents in the capacitor, I_C, and the coil, I_1, can be larger than the total current. The ratio of capacitor current to total current is the magnification or Q factor of the circuit.

Now $\quad I_C = \omega_0 C V \quad (\text{eqn (7.11)})$

$$I = \frac{V}{Z_D} \quad \text{from eqn (7.18)}$$

$$= \frac{CRV}{L}$$

hence $\quad Q = \dfrac{I_C}{I}$

$\quad\quad = \dfrac{\omega_0 CVL}{VCR}$

(7.19) $\quad\quad = \dfrac{\omega_0 L}{R}.$

This is very similar to eqn (7.7) for a series circuit. It should, however, be noted that the Q factor of a series circuit is *voltage* magnification and that of a parallel circuit *current* magnification. It should also be remembered that the impedance of a series circuit is *low* at resonance and that of a parallel circuit *high*. Hence *high* currents tend to flow in series resonant circuits, *low* currents in parallel ones. For this reason series circuits are often known as *acceptor* circuits (they accept current at the resonant frequency) and parallel ones as *rejector* circuits.

If the circuit considered above (a coil of 0.1 mH in parallel with a 100 pF capacitor) is connected to a 1 V supply, the total current, I, flowing if the coil has a resistance of 10 Ω is

$$I = \dfrac{1\,\text{V}}{100\,\text{k}\Omega} = 10\,\mu\text{A}.$$

The capacitor current is

$$\begin{aligned}I_C &= \omega_0 CV \\ &= 2\pi \times 1.591 \times 10^6 \times 100 \times 10^{-12} \times 1\,\text{A} \\ &= 1\,\text{mA}.\end{aligned}$$

and $\quad Q = \dfrac{1\,\text{mA}}{10\,\mu\text{A}} = 100.$

This could, of course, have been found thus

$$\begin{aligned}Q &= \dfrac{\omega_0 L}{R} \\ &= \dfrac{2\pi \times 1.591 \times 10^6 \times 0.1 \times 10^{-3}}{10} \\ &= 100.\end{aligned}$$

The current in the coil is found by calculating the impedance of the coil

$$\begin{aligned}Z &= \sqrt{(R^2 + \omega_0^2 L^2)} \\ &= \sqrt{[10^2 + (2\pi \times 1.591 \times 10^6 \times 0.1 \times 10^{-3})^2]} \\ &= 999.7\,\Omega.\end{aligned}$$

$$I_1 = \dfrac{1}{999.7}\,\text{A}$$

$\quad\quad = 1.0003\,\text{mA}$, almost the same as I_C.

Parallel resonance

Power factor correction

We have seen earlier than an alternator capable of supplying a current I at a voltage V will only supply a power VI if $\phi = 0$, i.e. if the power factor of the load is unity. If, at the other extreme, the power factor is zero, the alternator, although supplying its full output, is supplying no power and the supplying authority is receiving no income. Clearly, it is to the authority's advantage to supply loads which have a power factor of unity. Domestic appliances tend to have power factors of this order and present no problem. However, much industrial equipment has power factors lower than one. For example electric motors, used a great deal in industry, are inductive and have lagging power factors.

Consider a 400 V alternator with a maximum current output of 1000 A. It is capable of supplying 400 kW of power. If the consumer's load has a power factor of, say, 0.6, it only receives 400 × 0.6 or 240 kW. If the consumer is charged for the energy consumed, as is a domestic consumer, the authority only receives 60 per cent of the income which *could* be obtained from the generator. The generating authority thus encourages consumers to operate with power factors near to one. It is not necessary here to go into the full details of industrial tariffs, but, in effect, the consumer pays for VA rather than power. (In practice it is somewhat more complicated than this.)

Clearly a consumer can make the power factor of his equipment unity by adding a capacitor in parallel such that resonance is produced (or a coil if the equipment is capacitive). The required value can be found from eqn (7.16) or as follows. Consider the alternator referred to above. Let an inductive circuit with a power factor of 0.6 be taking the full 1000 A. The phasor diagram is shown in Fig. 7.21. To produce resonance requires a capacitor taking a current equal to the reactive, or vertical, component of I. This is

$$I \sin \phi = 1000 \sin 53.1°$$
$$= 800 \text{ A}.$$

The required reactance is $V/I = \frac{400}{800} = 0.5 \, \Omega$. If the frequency is 50 Hz the value of capacitance is

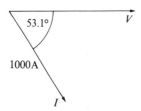

Fig. 7.21.

$$C = \frac{1}{2\pi f X_C}$$

$$= \frac{1}{2\pi \times 50 \times 0.5} \text{ F}$$

$$= 6366\,\mu\text{F}.$$

The total current now taken from the alternator is $I\cos\phi = 600$ A. In other words the alternator can now supply another 400 A to another consumer.

In practice, consumers do not usually correct the power factor of industrial equipment to unity for economic rather than electrical considerations. The capacitors involved cost money and need maintaining and the capital cost could be invested and bring in interest elsewhere.

Assume that in the problem above a correction to 0.9 lagging was the most economical. (Calculations to find the most economical power factor are beyond the scope of this book.) The phasor diagram (Fig. 7.22) shows the original current I and the total current I_1 (at an angle of $\cos^{-1} 0.9 = 25.8°$). Clearly, the length of the phasor I_C must equal the distance XY on the diagram, i.e. the difference between the vertical components of I and I_1. Now the horizontal component of I_1 must equal the horizontal component of I, as I_C has no horizontal component. This equals $1000 \cos 53.1°$ or 600 A. Hence

$$I_1 \cos 25.8° = 600$$

$$I_1 = \frac{600}{0.9}$$

$$= 666.7 \text{ A}.$$

The vertical component of I_1 is

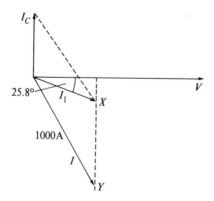

Fig. 7.22.

$I_1 \sin 25.8°$
$= 666.7 \sin 25.8° = 290.2$ A.

The vertical component of I is $1000 \sin 53.1 = 800$ A.
Hence

$$I_C = XY = 800 - 290.2$$
$$= 509.8 \text{ A}.$$

$$X_C = \frac{400}{509.8}$$
$$= 0.785 \text{ } \Omega$$

and $C = \dfrac{1}{2\pi \times 50 \times 0.785}$ F

$$= 4055 \text{ } \mu\text{F}.$$

The cost of the capacitors depends not only on their capacitance, but also on their VA rating, that is the product of voltage and current. In the example above,

$$VA = 400 \times 509.8$$
$$= 203\,920 \text{ VA}$$
$$= 203.92 \text{ kVA}.$$

Examples 7.4

1. A coil has a resistance of 15 Ω and an inductance of 0.2 H. It is in parallel with a 100 μF capacitor. Determine the frequency at which the circuit is purely resistive and the value of this resistance.

Solution

$$f_0 = \frac{1}{2\pi} \sqrt{\left(\frac{1}{LC} - \frac{R^2}{L^2}\right)}$$

$$= \frac{1}{2\pi} \sqrt{\left(\frac{1}{0.2 \times 100 \times 10^{-6}} - \frac{15^2}{0.2^2}\right)}$$

$$= \frac{1}{2\pi} \sqrt{(50\,000 - 5625)}$$

$$= \frac{210.65}{2\pi}$$

$$= 33.53 \text{ Hz}.$$

$$Z_D = \frac{L}{CR}$$

$$= \frac{0.2}{100 \times 10^{-6} \times 15}$$

$$= 133.3 \text{ } \Omega.$$

158 A.C. circuits II

2. If the circuit of question 1 is connected to a 200 V supply at the resonant frequency, determine the current in each component and the total current. What is the power factor of the coil?

Solution

$$X_C = \frac{1}{2\pi f_0 C}$$

$$= \frac{1}{2\pi \times 33.53 \times 100 \times 10^{-6}}$$

$$= 47.47 \, \Omega.$$

$$I_C = \frac{200}{47.47}$$

$$= 4.21 \, \text{A}.$$

$$Z_1 = \sqrt{(R^2 + \omega_0^2 L^2)}$$
$$= \sqrt{[15^2 + (2\pi \times 33.53 \times 0.2)^2]}$$
$$= \sqrt{(225 + 1775.4)}$$
$$= 44.73 \, \Omega.$$

$$I_1 = \frac{200}{44.73}$$

$$= 4.47 \, \text{A}.$$

$$\text{Total current}, I = \frac{200}{133.3}$$

$$= 1.5 \, \text{A}.$$

$$\phi_1 = \tan^{-1} \frac{2\pi \times 33.53 \times 0.2}{15}$$

$$= 70.4°.$$

$$\text{Power factor} = \cos 70.4°$$
$$= 0.34.$$

3. A coil of resistance 1000 Ω and inductance 0.5 H is in parallel with a capacitor, C. Find the value of C to produce resonance at 50 Hz.

Solution

$$f_0^2 = \frac{1}{4\pi^2}\left(\frac{1}{LC} - \frac{R^2}{L^2}\right)$$

$$4\pi^2 f_0^2 = \frac{1}{LC} - \frac{R^2}{L^2}$$

$$\frac{1}{LC} = 4\pi^2 f_0^2 + \frac{R^2}{L^2}$$

$$\frac{L}{C} = 4\pi^2 f_0^2 L^2 + R^2$$

$$\frac{L}{C} = Z_1^2$$

$$C = \frac{L}{Z_1^2} \quad \text{(this is, in fact, eqn (7.15))}.$$

$$Z_1^2 = R^2 + 4\pi^2 f_0^2 L^2$$
$$= 10\,000 + 4\pi^2 \times 50^2 \times 0.5^2$$
$$= 34\,674.$$

$$C = \frac{0.5}{34\,674}\,F$$

$$= 14.42\,\mu F.$$

4. A piece of industrial equipment takes a current of 100 A from a 400 V, 50 Hz supply at a power factor of 0.5 lagging. Find the capacitance required to increase the power factor to unity and state how it is connected. What current is now taken from the supply?

Solution

It is connected in parallel.

Fig. 7.23 shows a phasor diagram. Note that $\cos^{-1} 0.5$ is $60°$. The capacitor must take a current equal to the vertical component of the 100 A. This is

$$100 \sin 60°$$
$$= 86.6\,A$$
$$= I_C$$

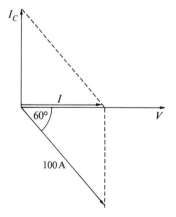

Fig. 7.23.

$$X_C = \frac{400}{86.6}$$
$$= 4.62 \, \Omega.$$
$$C = \frac{1}{2\pi \times 50 \times 4.62} \, F$$
$$= 689 \, \mu F.$$

The current now taken, I, is given by
$$I = 100 \cos 60$$
$$= 50 \, A.$$

5. A 400 V, 50 Hz, 20 h.p. motor has an efficiency of 85 per cent and a power factor of 0.87 lagging. Find the value of the capacitance required in parallel to increase the power factor to 0.95 lagging. What is the kVA rating of this capacitor?

Solution

$$\text{Input h.p.} = \frac{20}{0.85}$$
$$= 25.53 \, \text{h.p.}$$

This is $25.53 \times 746 \, W = 17\,553 \, W$.

$$P = V I_m \cos \phi \quad \text{where } I_m = \text{motor current}$$

$$I_m = \frac{17\,553}{400 \times 0.87}$$
$$= 50.44 \, A.$$

Now $\cos^{-1} 0.87 = 29.5°$ and
$$\cos^{-1} 0.95 = 18.2°.$$

Fig. 7.24 shows the phasor diagram.

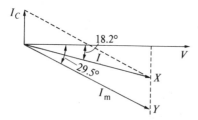

Fig. 7.24.

Horizontal component of I = horizontal component of I_m
$$= 50.44 \cos 29.5$$
$$= 43.9 \text{ A}.$$

$$I = \frac{43.9}{\cos 18.2}$$
$$= 46.21 \text{ A}.$$

Vertical component of I_m = $50.44 \sin 29.5$
$$= 24.84 \text{ A}.$$

Vertical component of I = $46.21 \sin 18.2$
$$= 14.43 \text{ A}.$$

$$XY = 24.84 - 14.43$$
$$= 10.41 \text{ A}.$$

$$X_C = \frac{400}{10.41}$$
$$= 38.42 \text{ }\Omega$$

$$C = \frac{1}{2\pi \times 50 \times 38.42} \text{ F}$$
$$= 82.85 \text{ }\mu\text{F}.$$

$$\text{kVA rating} = \frac{400 \times 10.41}{1000}$$
$$= 4.16 \text{ kVA}.$$

7D Test questions

1. A coil of resistance 2000 Ω and inductance 100 μH is in parallel with a 15 pF capacitor. Determine the resonant frequency.

2. If the circuit of question 1 is connected to a 2.5 V supply at the resonant frequency find
    (a) the dynamic impedance,
    (b) the total current in magnitude and phase,
    (c) the current in each component,
    (d) the power developed.

3. A 1500 pF capacitor is in parallel with a coil of inductance 1 mH and resistance 100 Ω. The circuit is connected to a supply of 100 V at the resonant frequency. Find
    (a) the resonant frequency,
    (b) the current in each component,
    (c) the total current,
    (d) the dynamic impedance,
    (e) the Q factor.

162 A.C. circuits II

4. A 50 Ω resistor and a perfect inductor of 0.15 H are in parallel across a 240 V, 50 Hz supply. Find
   (a) the current in each component,
   (b) the total current in magnitude and phase,
   (c) the power developed.

5. A 250 V, 50 Hz supply is connected to a 15 Ω resistor in parallel with a capacitor. If the total current is 25 A, find
   (a) the capacitance,
   (b) the frequency at which the total current will be 40 A.

6. A coil of inductance 0.05 H and resistance 20 Ω is connected to a 240 V, 50 Hz supply. It is found that the supply current does not alter when a certain capacitor is connected in parallel with the coil. Find the value of the capacitance.

7. A coil has a resistance of 70 Ω and an inductance of 100 μH. Find the value of the capacitance in parallel which will produce resonance at 340 kHz.

8. An inductive circuit takes a current of 460 A from a 500 V, 50 Hz supply at a power factor of 0.6. Find the capacitance required in parallel to improve the power factor to unity. What is the kVA rating of this capacitor?

9. A factory takes a power of 100 kW at a power factor of 0.55 lagging from a 440 V, 50 Hz supply. Find the capacitance and kVA rating of the capacitor required to increase the power factor to 0.9 lagging. What total current does the factory take before and after correction?

7.5 Three-phase supplies

The a.c. supplies of Chapters 6 and 7 have been what is known as *single phase*. That is the supply has two terminals with a sinusoidal waveform between them. Many alternators generate more than a single output, the outputs differing in phase. They are called *polyphase systems*, the main advantages being (i) a polyphase alternator produces more power than a single phase alternator of the same physical size, (ii) polyphase induction motors are self-starting, whereas single phase ones are not, (iii) polyphase induction motors have a higher efficiency and power factor than single phase ones, and (iv) there is a considerable saving in copper, a very expensive commodity, in transmission lines.

The only system which we shall consider is the one in general use for supplying power, the *three-phase system*. The alternator has, in effect, three coils at 120° to each other, and generates three waveforms that are identical but 120° apart in phase. Fig. 7.25 shows three such waveforms. The three waveforms are identified by colours—red, yellow, and blue. In the diagram, as yellow follows red and blue follows yellow, the phase sequence is said to be red, yellow, blue, or R Y B for short. If these three voltages are designated v_R, v_Y, and v_B and if their peak value is V_p then

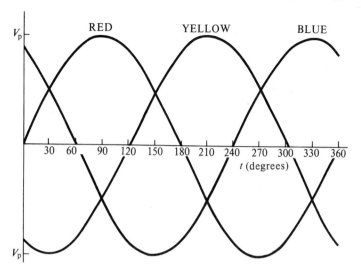

Fig. 7.25.

$$v_R = V_p \sin \omega t$$
$$v_Y = V_p \sin (\omega t - 120°) \qquad \text{eqn (7.20)}$$
$$v_B = V_p \sin (\omega t - 240°).$$

Fig. 7.26 shows a phasor diagram of a three phase supply. Note that as the phasors are rotating anticlockwise, v_Y is following v_R and v_B following v_Y. In this country V_p is about 340 V, meaning that the r.m.s. value of each waveform is $340/\sqrt{2}$ or about 240 V.

Obviously the three coils of a three-phase alternator could be used quite separately, each supplying its own load. It is more usual for them to be connected together in one of the ways to be described.

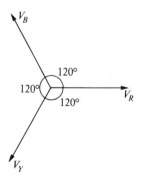

Fig. 7.26.

164   A.C. circuits II

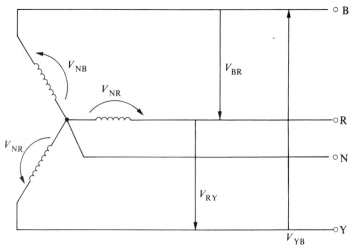

Fig. 7.27.

Star connection—four wire

In this arrangement one end of each coil is connected together. There are four output wires, one from the common or star point and the others from the other ends of each coil (Fig. 7.27). The wire from the common point is called the *neutral*, the others being the *lines*. The voltage across each coil is the voltage between the star point and the other end of the coil and is called the *phase voltage*. The three-phase voltages, all equal but 120° apart, are here designed V_{NR}, V_{NY}, and V_{NB}. Voltages also exist between the lines, called *line voltages* and are shown in Fig. 7.27 as V_{RY}, V_{YB}, and V_{BR}. Now a line voltage, such as V_{RY} is given by

(7.21) $$V_{RY} = V_{NY} - V_{NR}.$$

This can be seen from Fig. 7.27. If we travel from R to Y (i.e. through a voltage V_{RY}) we move from R to N *against* the direction of the arrow (hence the minus in eqn (7.21)) and from N to Y *with* the arrow. Now Fig. 7.28 shows V_{NY} and V_{NR} on a phasor diagram. To *subtract* V_{NR} from

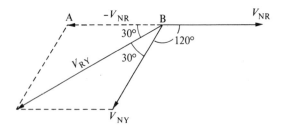

Fig. 7.28.

V_{NY} we turn V_{NR} through 180° and *add* it to V_{NY} as shown. V_{NB} has been omitted for clarity. As ABCD is a *rhombus* (a parallelogram with equal sides) the diagonals bisect at right angles (point E). From the triangle ABE, BE is AB cos 30°, that is $\sqrt{3}AB/2$. But BD is twice BE. Hence

$$BE = \sqrt{3}AB$$

(7.22) or $V_L = \sqrt{3}V_p$

where V_L = line voltage and V_p = phase voltage.

So the voltage between lines of a star-connected three-phase supply in this country is

$$V_L = \sqrt{3} \times 240 = 415.7 \text{ V r.m.s.}$$

Industrial consumers are supplied with this three-phase supply, that is with the three lines are neutral; whereas a domestic consumer receives one of the phases and neutral (240 V r.m.s.). It is conventional to refer to the line voltage of a three-phase supply unless otherwise stated.

Let us assume that a customer supplied with a three-phase supply has a load connected as a star, as shown in Fig. 7.29. This circuit is, in fact, like three separate single-phase systems, with a neutral current given, by Kirchhoffs First Law, as

(7.23) $$I_N = I_R + I_Y + I_B.$$

In many industrial applications the three loads of Fig. 7.29 are identical, for example the three coils of an induction motor. The load is then said to be balanced. In this case the line currents, I_R, I_Y, and I_B are equal in magnitude but 120° out of phase with one another (Fig. 7.30). Let the

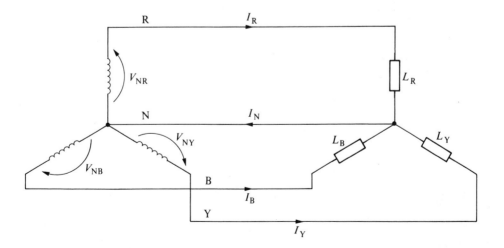

Fig. 7.29.

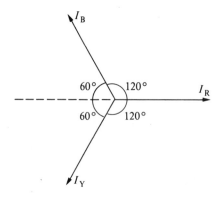

Fig. 7.30.

magnitude of each line current be I. To find the sum, I_N, we can resolve the currents into horizontal and vertical components. Resolving horizontally, I_Y gives $I \cos 60°$ to the left. This is $I/2$ to the left. Similarly I_B gives $I/2$ to the left. Thus the total horizontal component is I to the right (I_R) and I to the left (I_Y and I_B) which is zero.

Resolving vertically, I_R has no vertical component, I_Y gives $I \sin 60°$ downwards and $I_B \sin 60°$ upwards. Hence the total vertical component is also zero. The neutral current is zero meaning that the neutral wire may be omitted. It should be noted that the three line currents will never all be flowing in the same direction. It can be seen in Fig. 7.25 that two of the waveforms are always in the opposite direction to the third.

In domestic supplies, each phase is fed to a large area and it can be assumed that the currents taken on each phase will be about equal or that the load will be almost balanced. Of course the line currents may not be *quite* equal so that a small neutral current may flow. In practice the neutral wire can be made much thinner than the lines, leading to a saving of copper as referred to above.

It can be seen that the *phase currents*, that is the currents in the loads, are the same in a star-connected load as the *line currents*. The power in each load is

$$V_p I_L \cos \phi \quad \text{where } \cos \phi \text{ is the load power factor.}$$

Total power, $P = 3 V_p I_L \cos \phi$

and as $V_p = V_L/\sqrt{3}$, this is often written

(7.24) $$P = \sqrt{(3)} \, V_L I_L \cos \phi.$$

Summarizing, for a star-connected load

$$I_L = I_p$$
$$V_L = \sqrt{(3)} \, V_p$$
$$P = \sqrt{(3)} \, V_L I_L \cos \phi.$$

Three-wire star

A three-phase supply connected to a balanced star-connected load may have the neutral wire omitted, because the neutral carries no current. It then is a three-wire star system. The line and phase currents are still equal and eqns (7.22) and (7.24) still apply. It is worth noting that if the load is unbalanced, because the three loads are not identical or because one of the loads develops a fault (e.g. becomes open circuit), solutions for the line currents and the power become quite difficult.

Delta connection

Sometimes the three loads are connected as shown in Fig. 7.31. This method of connection is called *delta connection* owing to the similarity with a capital Greek delta, Δ. This method of connection is useful in some industrial applications because the loads have a higher voltage (the line voltage) across them. Of course, as far as the load is concerned, the line and phase voltages are now the same thing. Fig. 7.31 shows that now a line current is the difference between two phase currents, and a phasor diagram similar to Fig. 7.28 will give

(7.25) $$I_L = \sqrt{(3)} I_p.$$

The power in each load is

$$V_L I_p \cos \phi$$

and the total power

(7.26) $$P = 3 V_L I_p \cos \phi = \sqrt{(3)} V_L I_L \cos \phi.$$

Note that this is the same as eqn (7.24) for the power in a star-connected load. It is very important to remember that eqns (7.24) and (7.26) apply only to balanced loads.

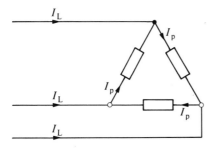

Fig. 7.31.

Examples 7.5

1. A 440 V, three-phase, 50 Hz supply is connected to a load consisting of three 100 Ω resistors. Find the line and phase currents and voltages if the resistors are connected (a) in star and (b) in delta.

Solution

(a) $V_L = 440$ V
$V_p = 440/\sqrt{3} = 254$ V.
$I_L = I_p = 254/100 = 2.54$ A.

(b) $V_L = V_p = 440$ V.
$I_p = 440/100 = 4.4$ A.
$I_L = 4.4 \sqrt{3} = 7.62$ A.

2. Find the total power in each case in question 1.

Solution

(a) $P = \sqrt{3} \times 440 \times 2.54 = 1936$ W.
(b) $P = \sqrt{3} \times 440 \times 7.62 = 5807$ W.

3. The three coils of a star-connected load have inductances of 0.05 H and resistances of 15 Ω. If they are connected to a 400 V, three-phase, 50 Hz supply, find the line currents and the total power.

Solution

$$Z = \sqrt{(15^2 + 4\pi^2 \times 50^2 \times 0.05^2)}$$
$$= \sqrt{(225 + 246.7)}$$
$$= 21.72 \, \Omega$$

$$\phi = \tan^{-1} \frac{2\pi \times 50 \times 0.05}{15}$$
$$= \tan^{-1} 1.047$$
$$= 46.3°.$$

$V_p = 400/\sqrt{3} = 231$ V.
$I_p = I_L = 231/21.72$
$= 10.6$ A.

$P = \sqrt{3} \times 400 \times 10.6 \times \cos 46.3°$
$= 5074$ W.

4. A 440 V, three-phase supply is connected to a resistive load. It is supplying 10 kW. Find the line and phase currents for (a) star and (b) delta connection.

Solution

(a) $\sqrt{3}\, V_L I_L = 10\,000$

$$I_p = I_L = \frac{10\,000}{\sqrt{3} \times 440}$$

$$= 13.1\,\text{A}.$$

(b) $I_L = 13.1$ A as before

$I_p = 13.1/\sqrt{3} = 7.56$ A.

7E Test questions

1. Three coils each of resistance 50 Ω and inductance 0.1 H are connected in star to a 440 V, 50 Hz supply. Calculate
    (a) the line current,
    (b) the power supplied, and
    (c) the power factor.

2. A delta-connected load has three loads consisting of a 50 Ω resistor in series with a 60 μF capacitor. If it is connected to a 440 V, 50 Hz supply find
    (a) the line and phase currents,
    (b) the total power.

3. A 400 V, three-phase, star-connected motor produces an output of 50 b.h.p. It has an efficiency of 0.85 and a power factor of 0.9. Find the line current.

4. A 400 V, 50 Hz, three-phase supply feeds a balanced, star-connected load. Each arm of the star consists of a 50 Ω resistor in parallel with a 75 μF capacitor. Find the line current and the total power.

8 Single-phase transformers

In Section 4.4 we were introduced to the principle of the transformer. If two coils are situated near to each other, a change of current in one, usually called the *primary*, causes an e.m.f. to be induced in the other, called the *secondary*. The e.m.f. in the secondary is proportional to the *rate of change* of current in the primary. If the primary current is a sine wave, the induced e.m.f. will also be sinusoidal, because the differential, or rate of change, of a sine wave is itself sinusoidal.

Transformers have many applications. As will be seen, they may be used to change the level of an alternating voltage. Thus a.c. generated at 11 000 V is stepped up to much higher voltages for transmission around the country. Other transformers are used to step it down to the required levels. Transformers are also used as impedance matching devices and as a means of isolation between two circuits.

Transformers are designed so that as much as possible of the flux produced by the primary links with the secondary. At power and audio frequencies the primary and secondaries are wound on top of one another, and are insulated from each other on a former or core of magnetic material (Fig. 8.1). The actual core construction and the type of cores used at higher frequencies will be dealt with later.

8.1 The perfect transformer

Power losses can occur in transformers in a number of ways. Power will be lost in the resistance of the windings and, as explained below, will also be lost in the core material. A perfect, or ideal, transformer is assumed to have no losses. Let us firstly consider such an ideal transformer on no load,

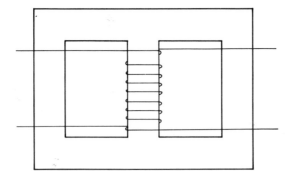

Fig. 8.1.

The perfect transformer

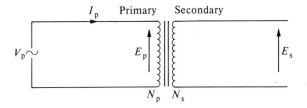

Fig. 8.2.

that is with the secondary open-circuited as shown in Fig. 8.2. This diagram also shows the circuit symbol for an iron-cored transformer. The primary winding has N_p turns and the secondary N_s. The *changing* current, I_p, in the primary will induce e.m.f.s E_p and E_s in the primary and secondary respectively. By Kirchhoff's Second Law, E_p must be just equal to V_p. This would not, of course, be the case if the winding resistance was considered. Hence the current I_p is just sufficient to produce a changing flux in the core which will make E_p equal to V_p. Its value will, in fact, equal V_p divided by the reactance of the primary. If there are no power losses, the current will be 90° out of phase with the applied voltage, because it is producing zero power. It will be in phase with the flux, ϕ, which it produces and it is called the magnetizing current, I_{mag}. Strictly speaking, in a perfect transformer the primary inductance is infinite and the magnetizing current is zero.

Now in a perfect transformer *all* the flux produced by the primary current links the secondary, to induce the secondary e.m.f. E_s. These two e.m.f.s will be proportional to the number of turns N_p and N_s. Hence

(8.1) $$E_p = k N_p$$
(8.2) $$E_s = k N_s$$

where k is a constant. (It can be shown that k is, in fact, $4.44 f \phi_{max}$ where f is the frequency and ϕ_{max} the peak value of the flux in the core.) Dividing eqn (8.1) by eqn (8.2)

(8.3) $$\frac{E_p}{E_s} = \frac{N_p}{N_s}.$$

Thus the secondary voltage is given by

(8.4) $$E_s = \frac{N_s}{N_p} \times E_p$$

and the transformer steps up or down voltage in the ratio of the number of turns.

Fig. 8.3 shows a phasor diagram of a perfect transformer on no load. Note the convention adopted in that E_p and E_s are in phase, although for

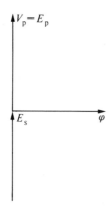

Fig. 8.3.

clarity E_s is drawn below E_p. E_p and E_s are shown, for convenience, as being about equal in magnitude.

Eqn (8.3) may be rewritten as

$$\frac{E_p}{N_p} = \frac{E_s}{N_s}. \tag{8.5}$$

E_p/N_p and E_s/N_s are the volts induced per turn, and are seen to be equal. Any coil wound on the same former will, in fact, have the same number of volts per turn.

If a resistive load is now placed across the secondary, the situation alters (Fig. 8.4). A secondary current I_s must flow, in phase with E_s, where

$$I_s = \frac{E_s}{R_L}. \tag{8.6}$$

This current will, of course, produce a flux in the core, proportional to $I_s N_s$, the secondary ampere-turns. But the flux in the core must remain as it was before the load was connected because, as explained earlier, it was just sufficient to produce the e.m.f. E_p equal to V_p, and V_p has not altered. The total flux can only remain constant if a primary current I_p flows to produce an equal amount of flux to cancel that produced by I_s, leaving the original flux as it was. (This current I_p is in addition to the magnetizing current.) Thus

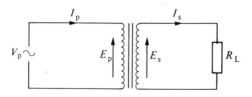

Fig. 8.4.

(8.7) $$I_p N_p = I_s N_s$$

(8.8) or $$\frac{I_p}{I_s} = \frac{N_s}{N_p}.$$

It is worth noting that as

$$E_s = \frac{N_s}{N_p} E_p \quad \text{(eqn (8.4))}$$

and $$I_s = \frac{N_p}{N_s} I_p \quad \text{(from eqn (8.8))}$$

(8.9) then $$E_s I_s = E_p I_p.$$

For a resistive load this indicates that the power output equals the power input. This, of course, follows from the fact that the perfect transformer has no power losses and indicates an efficiency of 100 per cent. Fig. 8.5 is a phasor diagram of a perfect transformer with a resistive load. Although I_p and I_s are in phase, Fig. 8.4 shows that they are flowing in opposite directions as far as the transformer core is concerned, and the fluxes they produce just neutralize one another.

Examples 8.1

1. A perfect transformer has a turns ratio of 100 to 50 (primary to secondary). If a voltage of 240 V, 50 Hz is applied to the primary find
(a) the secondary voltage,
(b) the secondary current for a resistive load of 25 Ω,
(c) the primary current in case (b).

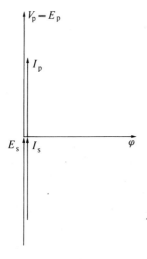

Fig. 8.5.

Solution

(a) $E_s = \dfrac{E_p N_s}{N_p}$

$= \dfrac{240 \times 50}{100}$

$= 120\,\text{V}.$

(b) $I_s = \dfrac{120}{25}$

$= 4.8\,\text{A}.$

(c) $I_p = \dfrac{I_s N_s}{N_p}$

$= \dfrac{4.8 \times 50}{100}$

$= 2.4\,\text{A}.$

2. An ideal transformer has 0.2 V/turn. The secondary load is a 40 Ω resistor. If the secondary current is 5 A when the primary voltage is 100 V find
(a) the primary current,
(b) the number of primary and secondary turns.

Solution

(a) $E_s = 5 \times 40$
$= 200\,\text{V},$

turns ratio $\dfrac{N_s}{N_p} = \dfrac{200}{100} = 2,$

$I_p = 5 \times 2 = 10\,\text{A}.$

(b) 0.2 V/turn ≡ 5 turns/V

$N_p = 100 \times 5 = 500$

$N_s = 200 \times 5 = 1000.$

3. A perfect transformer is designed to operate from the 240 V, 50 Hz mains. The primary has 1500 turns and the secondary 40 turns. If the secondary current is 2.4 A find the primary current and the secondary voltage.

Solution

$I_p = \dfrac{2.4 \times 40}{1500}$

$= 0.064\,\text{A},$

$$V_s = \frac{240 \times 40}{1500}$$
$$= 6.4 \text{ V}.$$

4. A perfect 1:10 step-up transformer has a 1000 Ω secondary load. If the primary voltage is 200 V find
(a) the secondary voltage and current,
(b) the primary current,
(c) the 'apparent' resistance seen at the primary terminals.

Solution

(a) $E_s = \dfrac{200 \times 10}{1}$

$\qquad = 2000 \text{ V}$

$\quad I_s = \dfrac{2000}{1000}$

$\qquad = 2 \text{ A}.$

(b) $I_p = \dfrac{2 \times 10}{1}$

$\qquad = 20 \text{ A}.$

(c) $R_{in} = \dfrac{200}{20}$

$\qquad = 10 \text{ Ω}.$

8.2 Matching

One of the circuit theorems dealt with in Chapter 2 was the maximum power theorem. This theorem applies also to a.c. circuits with resistive loads. (It is a little more complicated if the load includes any reactance.) The theorem states that a generator will deliver the maximum power to a load if the load resistance is equal to the generator resistance. Sometimes a generator and load have resistances of completely different orders of magnitude. Consider, for example, a generator of e.m.f. 10 V and internal resistance 2 kΩ feeding a 20 Ω load (Fig. 8.6). The current I is

$$I = \frac{10}{2020} \text{ A}$$
$$= 4.95 \text{ mA}$$

and the power in the load

$$P = I^2 R$$
$$= (4.95 \times 10^{-3})^2 \times 20 \text{ W}$$
$$= 0.49 \text{ mW}.$$

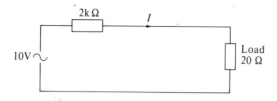

Fig. 8.6.

This is rather low when one considers the power which the generator would deliver to a load of $2\,\text{k}\Omega$. In this case the current would be $10/4000$ A or 2.5 mA and the power $(2.5 \times 10^{-3})^2 \times 2000$ W or 12.5 mW. There is, of course, no way in which this generator could supply more than 12.5 mW to a load. Example 4(c) above shows that a transformer may be used to alter the apparent resistance of a load. Consider Fig. 8.7.

The input resistance, R_{in}, is

$$R_{in} = \frac{E_p}{I_p}.$$

But $E_p = \dfrac{E_s N_p}{N_s}$ (from eqn (8.3))

hence $R_{in} = \dfrac{E_s N_p}{I_p N_s}.$

Further $I_p = \dfrac{I_s N_s}{N_p}$ (from eqn (8.8))

so that $R_{in} = \dfrac{E_s N_p N_p}{I_s N_s N_s}$

$$= \frac{E_s}{I_s}\left(\frac{N_p}{N_s}\right)^2$$

and $\dfrac{E_s}{I_s} = R_L$

(8.10) $R_{in} = \left(\dfrac{N_p}{N_s}\right)^2 R_L.$

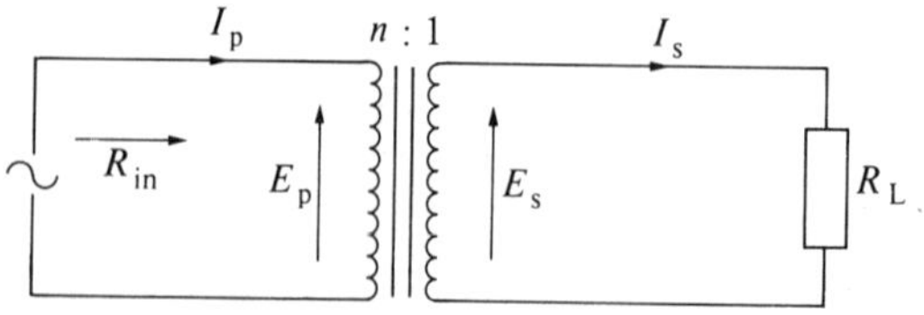

Fig. 8.7.

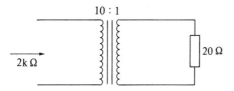

Fig. 8.8.

Thus R_L *appears* to be $(N_p/N_s)^2$ times greater at the transformer primary. If we connect the 20 Ω load of Fig. 8.6 to the secondary of a 10:1 step-down transformer (Fig. 8.8) it appears to be

$$\frac{10}{1}^2 \times 20 = 2000 \, \Omega$$

at the primary. Imagine a generator of, say, 1 V connected to the primary. The secondary voltage will be

$$1 \times \frac{1}{10} = 0.1 \, \text{V}$$

and the secondary current

$$I_s = \frac{0.1}{20} = 0.005 \, \text{A or 5 mA.}$$

The primary current is

$$I_p = \frac{5 \times 1}{10} = 0.5 \, \text{mA.}$$

So the primary receives 0.5 mA from 1 V and the secondary receives 5 mA at 0.1 V (0.5 mW in each case). The generator *thinks* it is feeding a load of 1 V/0.5 mA = 2000 Ω yet it is actually supplying power to a 20 Ω load.

If we connect the circuit of Fig. 8.8 to our 10 V, 2 kΩ generator considered earlier, the primary voltage must be 5 V (Fig. 8.9). The secondary voltage will be 0.5 V and the secondary current 0.025 A. The load power is

$$P = (0.025)^2 \times 20 \, \text{W}$$
$$= 12.5 \, \text{mW}$$

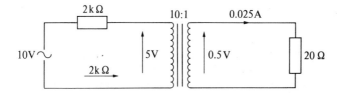

Fig. 8.9.

the maximum which the generator can supply. The load is said to be *matched* to the generator by the transformer. Transformers are used as much for matching as for their ability to step up or down voltages (although, of course, they are doing this as well). Question 1 in Example 8.2 gives an example of a low-resistance generator matched to a higher resistance load. Matching is required frequently in electronic engineering. Examples are the matching of generators such as microphones to transmission lines and the matching of valves and transistors to low-resistance loudspeakers.

Examples 8.2

1. A gramophone pick-up generates 0.5 mV and has an internal resistance of 5 Ω. It is required to feed an amplifier with an input resistance of 50 kΩ. Find
(a) the amplifier input power if it is connected directly to the amplifier,
(b) the input power if correctly matched,
(c) the turns ratio of the transformer required in (b).

Solution

(a) $\quad$ Current $= \dfrac{0.5 \text{ mV}}{50.005 \text{ k}\Omega} = 10^{-8}$ A.

Load power $= (10^{-8})^2 \times 50 \times 10^3$
$\qquad\qquad\;\; = 5 \times 10^{-12}$ W.

(b) If matched correctly, the load appears to be 5 Ω.

$$\text{Current} = \dfrac{0.5 \text{ mV}}{10} = 50 \,\mu\text{A}.$$

Load power $= (50 \times 10^{-6})^2 \times 5$
$\qquad\qquad\;\; = 1.25 \times 10^{-8}$ W.

(c) The transformer must be a step-up one of ratio

$$1 : \sqrt{\left(\dfrac{50\,000}{5}\right)}$$
$= 1 : \sqrt{10\,000}$
$= 1 : 100.$

2. A thermionic valve is feeding an 8 Ω loudspeaker via a 20:1 step-down transformer. If the valve acts as a 50 V (r.m.s.) a.c. generator with an internal resistance of 5 kΩ, determine
(a) the resistance looking into the transformer primary,
(b) the voltage across and current in the loudspeaker,
(c) the power developed in the loudspeaker.

Is the matching optimum? If not, what turns ratio transformer would be required?

Solution

(a) $\left(\frac{20}{1}\right)^2 \times 8 = 3200\,\Omega$

(b) $50 \times \frac{1}{20} = 2.5\,\text{V}$

$I = \frac{2.5}{8} = 0.313\,\text{A}$

(c) $I^2R = (0.313)^2 \times 8 = 0.78\,\text{W}.$

No. turns ratio required $= \sqrt{\left(\frac{5000}{8}\right)} = 25:1.$

8.3 Transformer losses

As explained earlier, power losses in a transformer can occur in two ways. *Iron losses* are the losses occurring in the magnetic material forming the core. *Copper losses* are the losses due to the resistance of the coils.

Iron losses

Power is lost in the core due to two effects. Firstly, the flux in the core is continually changing because the applied voltage is sinusoidal. The magnetic material is being taken round its hysteresis loop once during every cycle of the supply waveform. The power loss is clearly proportional to the area of the hysteresis loop and to the supply frequency. It can be reduced by using materials with narrow loops.

The second cause of power loss in the core is due to what are known as *eddy currents*. These are currents flowing round the core material due to the e.m.f.s induced in the core by the changing primary voltage. The core is itself acting as a secondary. This loss can be reduced by using material with a high resistivity and also by laminating the core. The core is built up of thin sheets or laminations (0.355 to 0.510 mm thick), insulated from one another. This greatly reduces the paths round which the eddy currents can flow, whilst hardly affecting the magnetic properties of the core. Even with laminations, however, iron losses become excessive at high frequencies. One solution is to use air-cored transformers which, obiously, have no iron losses. Another solution is to use cores made with certain ferrites. These materials have a very high magnetic permeability and also an exceedingly high electrical resistivity and are suitable up to microwave frequencies. These ferrite cores are largely replacing dust cores, in which the material consists of fine iron particles suspended in a non-conducting medium.

Copper losses

Currents flowing in the transformer coils produce power losses of I^2R in each coil. The only way in which these losses can be reduced is to reduce the resistance of the coils, which is often not practicable.

Open- and short-circuit tests

If the transformer is on no load, there will be no secondary current and the only primary current is that required to magnetize the core and overcome the iron losses. This no-load primary current is very small, and copper losses can be neglected as they are proportional to the current squared. Iron losses, however, depend on the primary applied voltage and are almost constant, unless the primary voltage is altered. If the input power is measured using a wattmeter, it will equal the value of the iron losses.

Whilst one does not normally put a short circuit across the secondary, a useful measurement can be made by doing just this. A very small primary voltage is applied and increased until the full-rated secondary current is flowing. The copper losses will then be at their full value, but the iron losses will be negligible because of the very small primary voltage. The input power is now a measure of the full copper losses. The copper losses on any other load can be calculated from the fact that they are proportional to I^2. It can be shown that the efficiency of a transformer is greatest when the load is such that the copper losses and iron losses are equal.

Examples 8.3

1. On no-load a 6600/400 V transformer takes a primary current of 0.8 A and the input power is 1.1 kW. Draw a phasor diagram showing this current and the applied 6600 V. What loss does the 1.1 kW represent?

Solution

$$VI = 6600 \times 0.8 = 5280 \text{ VA}.$$

$$\text{Power factor} = \frac{1100}{5280} = 0.21.$$

$$\phi = \cos^{-1} 0.21 = 77.9°.$$

The phasor diagram is shown in Fig. 8.10. The 1.1 kW represents the iron losses.

2. An ammeter (short circuit) is connected across the secondary of the transformer of question 1. The primary voltage is increased from zero until the full rated secondary current of 200 A is flowing. The wattmeter in the primary circuit now reads 1.4 kW. What does this represent? What primary

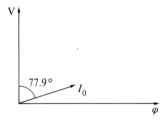

Fig. 8.10.

current is flowing? If the primary resistance is 0.5 Ω, find the resistance of the secondary.

Solution

The power is the copper loss.
The no load current is now zero because of the very low primary voltage.

$$I_p = \frac{400}{6600} \times 200 = 12.1 \text{ A}.$$

Copper loss in primary $= (12.1)^2 \times 0.5 = 73.2$ W.

Copper loss in secondary $= 1400 - 73.2 = 1326.8$ W.

$$R_s = \frac{1326.8}{200^2} = 0.033 \text{ Ω}.$$

3. (a) Calculate the iron and copper losses of the transformer of question 1 if operating at 6600 V on the primary but with only 100 A in the secondary.

   (b) At what secondary current will the efficiency of the transformer be a maximum?

     (c) If the power factor of the load is unity, what is the efficiency
        (i) at full load,
       (ii) at the load producing maximum efficiency?

Solution

   (a) The iron losses are still 1.1 kW.
As the currents are half their full value the copper losses will be a quarter of their full value,

i.e. $\dfrac{1.4}{4} = 0.35$ kW.

   (b) Maximum efficiency when copper losses are 1.1 kW. If this occurs at n times full load

$$n^2 \times 1.4 = 1.1$$

182 Single-phase transformers

$$n^2 = 0.786$$
$$n = 0.886,$$

Secondary current $= 0.886 \times 200$
$$= 177.3 \text{ A}.$$

(c) (i) Power $= 400 \times 200 = 80\,000$ W or 80 kW.

Input power $= 80 + 1.1 + 1.4 = 82.5$ kW.

$$\text{Efficiency} = \frac{80}{82.5}$$
$$= 0.9697 \quad \text{or } 96.97 \text{ per cent}.$$

(ii) Output $= 400 \times 177.3 = 70\,920$ W or 70.92 kW.

Input $= 70.92 + 1.1 + 1.1 = 73.12$ kW.

$$\text{Efficiency} = \frac{70.92}{73.12}$$
$$= 0.97 \quad \text{or } 97 \text{ per cent}.$$

4. If the secondary load is 200 A at unity power factor, determine, for the transformer of question 1, the full primary current when the primary voltage is 6600 V.

Solution

$I_0 = 0.8$ A at a power factor of 0.21 as before.

$I_s = 200$ A.

This causes a primary current, at unity power factor, of

$$200 \, \frac{400}{6600} = 12.1 \text{ A}.$$

The phasor diagram is shown in Fig. 8.11.

Fig. 8.11.

Total vertical = $12.1 + 0.8 \cos 77.9° = 12.27$ A.
Total horizontal = $0.8 \sin 77.9° = 0.78$ A.
Total current = $\sqrt{(12.27^2 + 0.78^2)}$
$= 12.29$ A.

8A Test questions

1. A 5:1 perfect transformer (step-down) is connected to the 240 V, 50 Hz mains. Find
    (a) the secondary voltage,
    (b) the primary and secondary currents if the secondary load is a 100 Ω resistor,
    (c) the primary and secondary currents if the secondary load is a coil of resistance 10 Ω and inductance 0.05 H,
    (d) the input power in cases (b) and (c).

2. An ideal transformer has a secondary voltage of 50 V when connected to a 240 V, 50 Hz supply. If the load is a 40 Ω resistor and the primary has 2000 turns find
    (a) the number of secondary turns,
    (b) the volts/turn,
    (c) the secondary current,
    (d) the primary current,
    (e) the input resistance at the primary.

3. A microphone has an internal resistance of 100 kΩ and generates 1.5 mV r.m.s. If it is connected directly to a line (cable) of input resistance 80 Ω, find the power at the line input. What turns ratio is required for a transformer which will produce correct matching, and what power would be developed at the line input using such a transformer?

4. A generator produces 100 V and has an internal resistance of 200 Ω. It is connected via a 1:4 step-up transformer to a resistive load. Find
    (a) the value of the load which will receive maximum power,
    (b) the primary voltage and current,
    (c) the secondary voltage and current,
    (d) the power in the load.

5. A transformer can deliver a maximum VA of 500 kVA. The open and full-current short circuit tests gave results of 3 kW and 4 kW respectively. Determine
    (a) the efficiency for a load of unity power factor at full output,
    (b) the efficiency for a load of power factor 0.7 at full output.

6. A transformer has 200 primary and 100 secondary turns. On open circuit the primary takes 1 A at a power factor of 0.2 lagging from a 240 V supply. If the secondary is now loaded so that it takes a current of 10 A at unity power factor, determine the total primary current and its power factor.

7. Repeat question 6 if the secondary current is 10 A at a power factor of 0.8 lagging.

8. A 40 kVA transformer gave readings of 500 W and 800 W respectively on the open and full-current short-circuit tests. Find
   (a) the load for maximum efficiency at a power factor of 0.8,
   (b) the value of this efficiency.

9 D. C. machines

We have already dealt with the principles of both d.c. generators and motors. Section 4.3 considered the force acting on a conductor carrying a current in a magnetic field, the magnitude being given by eqn (4.2) and the direction by Fleming's Left-Hand Rule. Section 4.4 showed that a conductor moving in a magnetic field has an e.m.f. induced in it, the magnitude and direction being given by eqn (4.6) and Fleming's Right-Hand Rule respectively. This chapter considers these basic principles applied to motors and generators, although the detailed construction of d.c. machines is not dealt with.

The d.c. machine applies the principles of Chapter 4 to rotational motion. A d.c. machine will generate an e.m.f. if the moving part, or armature, is rotated. Conversely, if an e.m.f. is applied to the armature it will rotate. We shall consider first the d.c. generator. It is important to realize that the motor, when rotating, still acts as a generator, the generated e.m.f., by Lenz's Law, opposing the applied e.m.f.

9.1 D.C. generators

Fig. 6.3 (p.92) showed that a coil rotating in a uniform magnetic field produces a sinusoidal waveform. Of course, there are problems in extracting a current from the rotating coil, and *slip rings* may be used (Fig. 9.1). These are conducting rings which rotate with the coil with stationary *carbon brushes* pressing against them

To generate direct current requires that the direction of the e.m.f. appearing between the brushes does not change direction twice every coil rotation. This may be achieved by the use of a *split-ring commutator*, usually simply called a commutator (Fig. 9.2). The ring is now made in

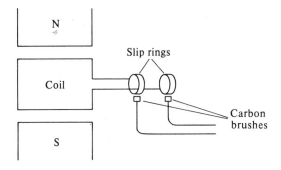

Fig. 9.1.

186 D.C. machines

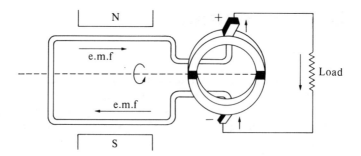

Fig. 9.2.

two halves, insulated from one another. If the coil is rotating so that the top is coming out of the paper the induced e.m.f. will be in the direction shown. Fig. 9.3 shows an end view of the commutator at three different points on the coils rotation. At (a) the induced e.m.f. is a maximum with coil side 'X' positive and the current flows in the direction shown. At (b)

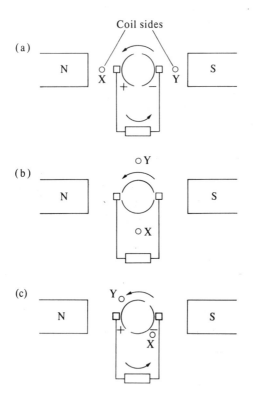

Fig. 9.3.

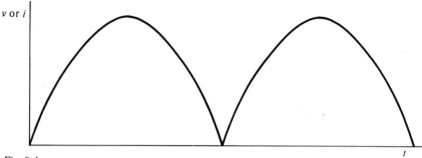

Fig. 9.4.

the e.m.f. is momentarily zero (the coil sides moving parallel to the flux), and at (c) the induced e.m.f. is between zero and maximum with coil side 'Y' positive. However, the commutator has ensured that the current is flowing in the same direction as at (a). The voltage, or current, waveform is shown in Fig. 9.4.

A real generator will have more than the two poles shown in Fig. 9.3, and will also have many coils wound on the armature. The armature will consist of laminations of magnetic material, with the coils wound in slots so that the air gaps in the magnetic circuit are as small as possible. The commutator will now contain many segments, each coil being connected to a pair of segments. The magnetic poles will, in all but very small machines, be produced by current flowing in field windings, as shown in Fig. 9.5, a four-pole machine. The magnetic field is through the pole, the

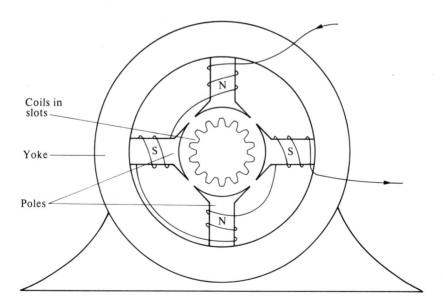

Fig. 9.5.

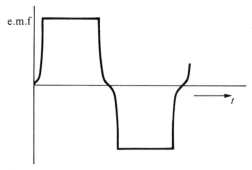

Fig. 9.6.

armature, and the *yoke*. Because of the number of poles and, to some extent, the shaping of the poles, the resultant e.m.f. is rather different to that shown in Fig. 9.4. Fig. 9.6 shows the (alternating) e.m.f. induced in one coil. The total e.m.f. between the brushes will be almost constant d.c.

It is not possible in this book to go into the details of the way in which the coils are connected together. It is sufficient to mention the two main methods.

Lap windings

The coils are arranged in a series–parallel combination such that there are as many parallel paths as there are poles on the machine. If the machine has a total number of conductors Z and if it has $2p$ poles (a number which must be even) then each of the parallel paths has $Z/2p$ conductors in series. It is, of course, the number in series which determines the e.m.f. between the brushes.

Wave windings

In a wave-wound machine there are two parallel paths independent of the number of machine may have. Hence the number of conductors in series in each parallel path is $Z/2$.

Generated e.m.f.

Each conductor, as it rotates, has an e.m.f. induced in it. In one rotation it passes $2p$ poles, and if each pole produces a useful flux of ϕ webers, each conductor will cut $2p\phi$ webers of flux per rotation. If the armature rotates at N rev/min, or $N/60$ rev/s, the time for one rotation is $60/N$ s. As the flux cut per second is equal to the induced e.m.f.

$$\text{e.m.f. per conductor} = \frac{2p\phi}{60/N}$$

$$= \frac{2p\phi N}{60} \text{ V}.$$

(9.1)

D.C. generators 189

If there are c parallel paths between the brushes ($c = 2p$ for a lap winding and 2 for a wave winding), then the number of conductors in series is Z/c and

(9.2) $$\text{generated e.m.f.} = \frac{Z}{c} \frac{2p\phi N}{60}.$$

In general, as Z, p, and c are constant for a given machine,

(9.3) $$\text{e.m.f.}, E = k\phi N$$

(9.4) or $E \propto \phi N$.

Terminal voltage

The above equations refer to the generated e.m.f., or the voltage across the terminals of the machine on open circuit. If the generator is supplying a current the terminal voltage will be less than E because of the voltage dropped in the resistance of the armature windings. If the armature resistance is R_a and if the generator is supplying a current I_a, the terminal voltage V must be

(9.5) $$V = E - I_a R_a.$$

Field current

The current for the field coils can either be produced by a separate d.c. supply (*separately excited generator*) or by the generator itself (*self-excited generator*). The self-excited generator depends on residual magnetism to start, and the field coils can either be in shunt (parallel) or series with the armature coils. If in shunt, the coils consist of a large number of turns of a relatively high resistance, i.e. low current, and if in series, hence taking the load current, they consist of just a few turns of as low a resistance as possible. It is, of course, the ampere-turns which matter as far flux is concerned.

Examples 9.1

1. A four-pole lap-wound generator has a useful flux per pole of 100 mWb. If there are 800 conductors in the armature, and the speed of rotation is 300 rev/min, find the generated e.m.f.

Solution

$$c = 2p = 4$$

$$\text{e.m.f.} = \frac{Z \times 2 \times p \times \phi \times N}{60c}$$

$$= \frac{800 \times 2 \times 2 \times 100 \times 10^{-3} \times 300}{60 \times 4}$$

$$= 400\,\text{V}.$$

2. A six-pole wave-wound generator develops 500 V on open circuit. If it has 400 conductors and a useful flux per pole of 50 mWb, calculate the speed of rotation.

Solution

$$N = \frac{60\, E\, c}{2p Z \phi}$$
$$= \frac{60 \times 500 \times 2}{2 \times 2 \times 50 \times 10^{-3} \times 400}$$
$$= 750 \text{ rev/min.}$$

3. The generator of question 2 has an armature resistance of 0.2 Ω and is supplying 100 A. Find the terminal voltage.

Solution

$$V = E - I_a R_a$$
$$= 500 - 0.2 \times 100$$
$$= 480 \text{ V.}$$

9A Test questions

1. Sketch a simple diagram of a four-pole d.c. machine showing the poles, field windings, armature, and yoke.

2. Explain the differences between slip rings and a commutator.

3. Explain, briefly, the difference between lap and wave windings. If a machine with $2p$ poles has a total number of conductors Z, how many conductors are in series in each parallel path for
    (a) lap and
    (b) wave winding?

4. A four-pole generator has 1000 conductors and rotates at 750 rev/min. If the flux per pole is 80 mWb, find the generated e.m.f. if the machine is
    (a) wave wound and
    (b) lap wound.

5. If the armature resistance of the generator in question 4 is 0.1 Ω, find the terminal voltage if the load current is 150 A for the lap-wound machine.

6. A d.c. generator generates an e.m.f. of 500 V on open circuit when running at 600 rev/min. The flux per pole is 100 mWb. Find the generated e.m.f. if
    (a) the speed is increased to 650 rev/min (flux constant),
    (b) the flux per pole is reduced to 80 mWb (speed constant).

7. If the speed of the machine in question 6 is increased to 700 rev/min, to what value must the flux per pole be changed if the generated e.m.f. is to remain at 500 V?

8. A d.c. generator is supplying a constant load current of 200 A. It has an armature resistance of $0.2\,\Omega$, and has a terminal voltage of 400 V when running at 800 rev/min. Find the terminal voltage if the speed is increased to 850 rev/min. Assume that the flux is constant.

9.2 The d.c. motor

If the generator described above has an e.m.f. applied to its brushes, the armature will rotate (if free to do so). As it rotates, an e.m.f. will be generated in it. It is acting as a generator, as well as a motor. The e.m.f. given by the equations above will be generated in a rotating armature, whatever is causing it to rotate. Of course, Lenz's Law tells us that the generated e.m.f. will oppose the applied voltage V. It is often called a *back e.m.f.* The net voltage, $V - E$, produces a current I_a in the armature windings. If the armature resistance is R_a,

(9.6) $$I_a = \frac{V - E}{R_a}.$$

This equation is more often written

(9.7) $$V = E + I_a R_a.$$

Before the armature starts to rotate, E will be zero and

(9.8) $$I_a = \frac{V}{R_a}.$$

This starting current may be very large and it will need limiting in all but the smallest motors. This subject is dealt with below. When running at speed, E might be very nearly equal to V, and the armature current quite small.

Speed

Now $E = k\phi N$. (eqn (9.3)).

Substituting in eqn (9.7)

$$V = k\phi N + I_a R_a$$

(9.9) and $$N = \frac{V - I_a R_a}{k\phi}.$$

In many practical cases V is very much larger than $I_a R_a$, and to a first approximation

(9.10) $$N \simeq \frac{V}{k\phi}$$

(9.11) or $$N \propto \frac{V}{\phi}.$$

192 D.C. machines

Note that for a given applied voltage

(9.12) $$N \propto \frac{1}{\phi}.$$

Torque

It is very important to know the torque which is produced by a motor. If the torque developed, in newton-metres is T then

$$\text{mechanical power developed} = \frac{2\pi NT}{60} \text{ W}.$$

Now if eqn (9.7) is written

$$V - I_a R_a = E$$

and multiplied throughout by I_a

(9.13) $$V I_a - I_a^2 R_a = E I_a.$$

$V I_a$ is obviously the input power and $I_a^2 R_a$ is the power lost, as heat, in the armature windings. Hence the power available for conversion to mechanical power is $E I_a$ and

$$E I_a = \frac{2\pi NT}{60}$$

(9.14) or $$T = \frac{60 E I_a}{2\pi N} \text{ newton-metres}.$$

But $$E = \frac{Z \times 2p \phi N}{60 c} \quad \text{(eqn (9.2))}$$

$$T = \frac{60 \times 2p \phi N Z I_a}{60 c \times 2\pi N}$$

$$= \frac{1}{\pi} I_a \frac{Z}{c} \phi p$$

(9.15) $$= 0.318 \frac{Z}{c} p \phi I_a.$$

For a given machine

(9.16) $$T \propto \phi I_a.$$

Examples 9.2

1. A motor takes 20 A from a 200 V supply. If the armature resistance is 0.6 Ω find the back e.m.f.

Solution

$$E = V - I_a R_a$$
$$= 200 - 12 = 188 \text{ V}.$$

2. The motor of question 1 is a six-pole, lap-wound motor with 500 conductors and a flux per pole of 50 mWb. If the armature current is 20 A, as in question 1, find the speed and torque.

Solution

$$\text{Generated e.m.f.} = \frac{2p\phi NZ}{60c} \quad \text{(eqn (9.2))}$$

$$= \frac{2 \times 3 \times 50 \times 10^{-3} \times N \times 500}{60 \times 2 \times 3}$$

$$= 0.417 \, N.$$

$$N = \frac{E}{0.417}$$

$$= \frac{188}{0.417}$$

$$= 451 \text{ rev/min}.$$

$$T = 0.318 \frac{Z}{c} p \phi I_a$$

$$= 0.318 \, \frac{500}{6} \, 3 \times 50 \times 10^{-3} \times 20$$

$$= 79.5 \, \text{Nm}.$$

9.3 Shunt motors

In a shunt motor the magnetic field is produced by coils which are wound in shunt, or parallel, with the armature coils. This is shown in Fig. 9.7. This diagram shows the armature back e.m.f. E and resistance R_a. The field windings, R_f, are in parallel with the armature and the terminal voltage is

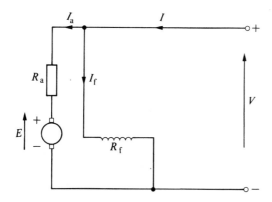

Fig. 9.7.

V. The total input current I comprises the armature and field currents, and

(9.17)
$$I = I_a + I_f.$$

As the flux depends on the ampere-turns, the field winding is designed to consist of a large number of turns and takes a current small compared with the armature current. This field current, V/R_f, is obviously constant for a given applied voltage. The speed of the motor is given by eqn (9.9)

$$N = \frac{V - I_a R_a}{k\phi}.$$

As explained above

$$N \simeq \frac{V}{k\phi}$$

and k and V are usually constant. But if V is constant, so is I_f and hence ϕ, so that the shunt motor can be seen to run at an approximately constant speed. This is a useful property in many applications. A graph of speed against torque for a shunt motor is shown in Fig. 9.8. Of course sometimes it is required to vary the speed of such a motor and as k and V are normally constant, this can best be done by altering either R_a or ϕ (eqn (9.9)). I_a, of course, depends on the load. R_a can, in effect, be increased by adding resistance in series with the armature (Fig. 9.9). The flux produced by the field current is not altered and increasing R will, from eqn (9.9), decrease the speed of the motor. Unfortunately the large armature current flowing in R wastes power. Further, the reduction in speed obtained depends on the load, the motor becoming less of a constant-speed machine, the speed falling more with torque than it does without the added resistance. However, the method is used, and can produce speeds down to zero.

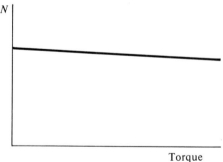

Fig. 9.8.

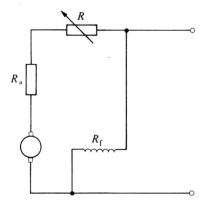

Fig. 9.9.

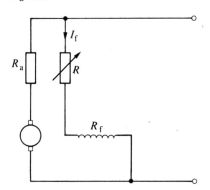

Fig. 9.10.

The flux can be altered by varying the field current (Fig. 9.10). It can sometimes be assumed, and will be for all problems in this book, that $\phi \propto I_f$. Increasing the added resistance will decrease the flux and, from eqn (9.9), increase the speed. This method of speed control has the advantages that because of the low field current there is not much power wasted in the added resistance and also that the motor will still act as a constant-speed motor at, of course, its new speed. However, the speed cannot be decreased by this method.

Obviously shunt motors are used where constancy of speed is required.

Examples 9.3

1. A d.c. shunt motor is running at 700 rev/min and taking 30 A from a 250 V supply. The armature resistance is 0.2 Ω. If the field current is reduced from 2 A to 1.9 A, find
(a) the new value of the supply current,
(b) the original and new values of the back e.m.f.
(c) the new speed.

196   D.C. machines

Solution

Initial armature current $I_{a1} = 30 - 2 = 28$ A.

Initial back e.m.f. $= E_1 = 250 - 28 \times 0.2$
$= 244.4$ V.

Now
$$I_{a1} \phi_1 = I_{a2} \phi_2$$
$$28 \times 2 = I_{a2} \times 1.9 \quad (\text{because } \phi \propto I_f)$$
$$I_{a2} = 29.5 \text{ A.}$$

Total new current $= 29.5 + 1.9 = 31.4$ A.

$$E_2 = 250 - 29.5 \times 0.2$$
$$= 244.1 \text{ V.}$$

$$N \propto \frac{1}{\phi}$$

$$\frac{N_1}{N_2} = \frac{\phi_2}{\phi_1} = \frac{1.9}{2}$$

$$N_2 = \frac{700 \times 2}{1.9}$$
$$= 736.8 \text{ rev/min.}$$

2. A 400 V d.c. shunt motor runs at 600 rev/min without load. The field resistance is 200 Ω. Find the extra resistance needed in series with the field to increase the no load speed to 800 rev/min. Assume that the flux is proportional to the field current.

Solution

$$I_{f1} = \frac{400}{200} = 2 \text{ A.}$$

$$\frac{N_1}{N_2} = \frac{\phi_2}{\phi_1} = \frac{I_{f2}}{I_{f1}}$$

(neglecting $I_a R_a$ as the motor is on no load).

$$\frac{600}{800} = \frac{I_{f2}}{2}$$

$$I_{f2} = \frac{1200}{800} = 1.5 \text{ A.}$$

$$R_f = \frac{400}{1.5} = 267 \, \Omega.$$

Extra resistance $= 67 \, \Omega$.

Shunt motors

3. A 400 V d.c. shunt motor has an armature resistance of 0.1 Ω and a field resistance of 200 Ω. When driving a constant torque load at 500 rev/min the armature current is 200 A. A resistor of 50 Ω is added in series with the field. Find the speed at which the motor now runs.

Solution

$$T \propto \phi I_a \quad \text{(eqn (9.16))}$$

$$\frac{T_1}{T_2} = \frac{\phi_1 I_{a1}}{\phi_2 I_{a2}}.$$

But $\quad T_1 = T_2$

$$\frac{\phi_1 I_{a1}}{\phi_2 I_{a2}} = 1$$

$$I_{a2} = I_{a1} \frac{\phi_1}{\phi_2} = I_{a1} \cdot \frac{I_{f1}}{I_{f2}}$$

$$I_{f1} = \frac{400}{200} = 2 \text{ A}$$

$$I_{f2} = \frac{400}{250} = 1.6 \text{ A}$$

$$I_{a2} = \frac{200 \times 2}{1.6} = 250 \text{ A}.$$

Also $\quad \dfrac{N_1}{N_2} = \dfrac{V - I_{a1} R_a}{V - I_{a2} R_a} \cdot \dfrac{I_{f2}}{I_{f1}}$

$$\frac{500}{N_2} = \frac{400 - 200 \times 0.1}{400 - 250 \times 0.1} \cdot \frac{1.6}{2}$$

$$= 0.81$$

$$N_2 = \frac{500}{0.81} = 617.3 \text{ rev/min}.$$

4. A 200 V d.c. shunt motor takes an armature current of 50 A and has an armature resistance of 0.1 Ω. It runs at 1000 rev/min with a certain load. Find the new speed if the load torque doubles.

Solution

$$T \propto \phi I_a$$

$$\frac{T_1}{T_2} = \frac{\phi_1 I_{a1}}{\phi_2 I_{a2}}$$

$$\frac{1}{2} = \frac{50}{I_{a2}} \quad (\phi \text{ constant})$$

$$I_{a2} = 100 \text{ A}.$$

198 D.C. machines

$$\text{Original back e.m.f.} = E_1 = 200 - 50 \times 0.1$$
$$= 195 \text{ V}.$$

$$E_2 = 200 - 100 \times 0.1$$
$$= 190 \text{ V}.$$

$$\frac{N_1}{N_2} = \frac{E_1}{E_2} \quad \text{(flux constant)}$$

$$N_2 = \frac{1000 \times 190}{195}$$
$$= 974 \text{ rev/min}.$$

5. A 400 V d.c. shunt motor runs at 850 rev/min with an armature current of 40 A. The armature resistance is 0.3 Ω. Calculate the resistance to be added in series with the armature to reduce the speed to 650 rev/min, assuming that the load torque is constant.

Solution

$$T \propto \phi I_a.$$

As T and ϕ are constant, I_a remains at 40 A.

$$E_1 = 400 - 40 \times 0.3$$
$$= 388 \text{ V}.$$

But
$$E \propto N \text{ (as } \phi \text{ is constant)}$$

$$\frac{E_1}{E_2} = \frac{N_1}{N_2}$$

$$E_2 = \frac{388 \times 650}{850}$$
$$= 296.7 \text{ V}.$$

$$296.7 = 400 - 40 R_a$$
$$40 R_a = 400 - 296.7$$
$$R_a = 2.58 \text{ Ω}.$$

$$\text{Extra resistance required} = 2.58 - 0.3$$
$$= 2.28 \text{ Ω}.$$

9B Test questions

1. An eight-pole, lap-wound motor has 700 conductors and a flux per pole of 75 mWb. If the armature current is 25 A, the armature resistance is 0.3 Ω, and the applied voltage is 250 V find
   (a) the back e.m.f.

(b) the speed,
(c) the torque.

2. A d.c. shunt motor takes an armature current of 150 A from a 500 V supply when running at 750 rev/min. It has an armature resistance of 0.2 Ω. Find the speed if the load torque is halved.

3. A d.c. shunt motor takes 20 A from a 200 V supply when running at 1000 rev/min. The armature and field resistances are 0.15 Ω and 100 Ω respectively. If the load torque remains constant calculate the speed at which it will run if a 33.3 Ω resistor is added in series with the field. Assume that the flux per pole is proportional to the field current.

4. Repeat question 3 if the field resistance is kept constant at 100 Ω, but a 0.1 Ω resistor is added in series with the armature.

5. If, in question 3, both resistors (33.3 Ω in series with the field and 0.1 Ω in series with the armature) are added at the same time, find the speed assuming that the load torque remains constant.

6. A 500 V d.c. shunt motor runs at 750 rev/min with an armature current of 50 A. The armature resistance is 0.4 Ω. If the load torque increases to one and a half times its previous value, find the resistance to be added in series with the armature to reduce the speed to 700 rev/min.

9.4 The series motor

In this case the field windings are in series with the armature and take the full armature current. Because this current might be quite large, the windings consist of a few turns of low resistance. Fig. 9.11 shows a series motor. Note that

(9.18) $$I_a = I_f.$$

Eqn (9.7) now becomes

(9.19) $$V = E + I_a(R_a + R_f)$$

and eqn (9.9) becomes

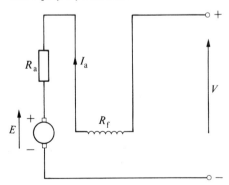

Fig. 9.11.

(9.20) $$N = \frac{V - I_a(R_a + R_f)}{k\phi}.$$

Again, if V is constant and if the volts drop, this time in the armature and field, is neglected

$$N \propto \frac{1}{\phi}.$$

However, in this motor the flux is produced by the armature current, and if it is assumed proportional to it

(9.21) $$N \propto \frac{1}{I_a}.$$

At low armature currents, corresponding to small loads, the speed may be very high, perhaps dangerously so. Such motors are usually used where there is always some load. They are unsuitable for applications where there is any danger of the load becoming disconnected, such as in the case of belt drives, where the belt might break.

The torque is given by eqn (9.16)

$$T \propto \phi I_a$$

but as $\phi \propto I_a$ it becomes

(9.22) $$T \propto I_a^2$$

which is fairly accurate unless saturation occurs. The equation quoted above enables a speed/torque curve to be drawn (Fig. 9.12). A curve of this type with a very high torque at low speeds is ideal for traction purposes, avoiding the need for a gear box, a device used to provide high starting torques in engines such as internal combustion engines.

The speed of a series motor may be controlled in various ways. The applied voltage is sometimes altered in traction applications. Like the

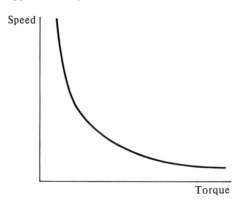

Fig. 9.12.

shunt motor, resistance may be added in series with the armature (and field in this case) with the same disadvantages. The field current can be altered by placing a diverting resistance in parallel with the field winding, thus reducing the field current and increasing the motor speed.

Examples 9.4

1. A 400 V d.c. series motor takes 100 A and runs at 800 rev/min. The total resistance of armature and field is 0.15 Ω. Find the current and speed if the load torque is halved. (Assume $\phi \propto I_a$.)

Solution

$$T \propto I_a^2$$

$$\frac{T_1}{T_2} = \frac{I_{a1}^2}{I_{a2}^2}$$

$$\frac{2}{1} = \frac{100^2}{I_{a2}^2}$$

$$I_{a2}^2 = \frac{100^2}{2}$$

$$I_{a2} = 70.7 \text{ A}.$$

$$\frac{N_1}{N_2} = \frac{V - I_{a1}(R_a + R_f)}{V - I_{a2}(R_a + R_f)} \cdot \frac{I_{a2}}{I_{a1}}$$

$$\frac{800}{N_2} = \frac{400 - 100 \times 0.15}{400 - 70.7 \times 0.15} \cdot \frac{70.7}{100}$$

$$= 0.7$$

$$N_2 = \frac{800}{0.7} = 1143 \text{ rev/min}.$$

2. A 200 V d.c. series motor has armature and field resistances of 0.15 Ω and 0.05 Ω respectively. It takes 100 A and has a speed of 900 rev/min when driving a constant torque load. What value of resistance is required in series with it to reduce the speed to 750 rev/min? (Assume $\phi \propto I_a$.)

Solution

$$\frac{T_1}{T_2} = \frac{I_{a1}^2}{I_{a2}^2} = 1$$

$$I_{a1} = I_{a2}.$$

$$\frac{N_1}{N_2} = \frac{V - I_{a1}(R_a + R_f)}{V - I_{a2}(R_a + R_f + R)} \cdot \frac{I_{a2}}{I_{a1}}$$

where R is the added resistance

202   D.C. machines

$$\frac{900}{750} = \frac{200 - 100 \times 0.2}{(200 - 100(0.2 + R))} \times \frac{100}{100}$$

$$1.2 = \frac{180}{200 - 100\,(0.2 + R)}$$

$$200 - 100\,(0.2 + R) = \frac{180}{1.2} = 150$$

$$100\,(0.2 + R) = 50$$
$$0.2 + R = 0.5$$
$$R = 0.3\,\Omega.$$

9.5 Motor starters

Question 1 in Examples 9.2 dealt with a 200 V motor taking 20 A with an armature resistance of 0.6 Ω. The back e.m.f. was found to be 188 V. This means that the net forward e.m.f. is 12 V, giving 20 A in 0.6 Ω. The back e.m.f. is produced because the armature is rotating in a magnetic field. When the motor is first switched on there can be no back e.m.f. and the starting current is

$$I = \frac{200}{0.6} = 333 \text{ A}.$$

A motor designed to operate with currents in the order of 20 A might well burn out with a current of this magnitude, even for a short period of time.

It is not intended here to discuss motor starters in detail, but only in principle. When a d.c. motor is first switched on a resistance is placed in series with it to limit the starting current. As the starting handle is slowly moved to the 'ON' position this resistance is gradually reduced until it becomes zero. The motor speeds up during this period, and the back e.m.f. increases. To limit the starting current in the above example to, say 40 A, would require a resistance of 200/40 or 5 Ω. The starting resistance would thus need to be 4.4 Ω.

9.6 Losses

There are many causes of loss of energy in a d.c. machine. These may be listed as

(a) Mechanical losses: these comprise friction and wind resistance.
(b) Copper losses: heat is lost in the armature and field coils.
(c) Iron losses: iron losses will occur in the armature because it is moving past magnetic poles, inducing eddy currents and subjecting the armature material to a hysteresis cycle. Iron losses may be reduced, as usual, by laminating the armature and by using a material with a small area hysteresis loop.

The efficiency of a machine is given by

(9.23) $$\text{efficiency} \frac{\text{input power} - \text{losses}}{\text{input power}}.$$

9C Test questions

1. A 500 V d.c. series motor runs at 750 rev/min when taking a current of 75 A. The total resistance, armature plus field, is 0.2 Ω. Find the speed and current if the load torque is doubled.

2. A 450 V d.c. series motor has a total resistance of 0.2 Ω and takes 150 A at a speed of 700 rev/min when driving a load. If the load torque is halved find the value of the resistance to be added in series with the motor to maintain the speed constant at 700 rev/min.

3. A 400 V d.c. shunt motor running on full load takes an armature current of 100 A and develops a back e.m.f. of 380 V. Determine the series resistance needed to limit the starting current to 10 A.

10 Electrical measurements

This chapter deals with the principles of certain measuring instruments and certain measuring techniques. It should be supplemented by practical work in the laboratory, for which there can be no substitute.

The first few sections are concerned with pointer or deflectional instruments, that is instruments in which a pointer moves across a scale to indicate the value of the quantity being measured. There are three forces involved in a deflection instrument. Firstly there must be a *deflecting torque* which is a function of the quantity being measured. It is the force or torque which moves the pointer. To obtain a measurement it must be balanced by a *controlling torque* in the opposite direction, the pointer coming to rest when the two torques are equal (and opposite). Ideally the controlling torque will be proportional to the deflection, and it is almost always provided by a coil spring. Lastly there will be a *damping torque*, the purpose of which is to prevent undue oscillations of the needle about its final position. It is a torque which should only be present when the pointer is moving. Fig. 10.1 shows the response of an instrument pointer as a function of time. Not enough damping (the underdamped condition) will result in oscillations about the final position. If overdamped, the movement is too sluggish. Critical damping is the borderline between the pointer oscillating and not oscillating.

10.1 Measurement of current

One ideal of any measuring instrument is that its use does not alter the quantity being measured. Clearly then, an instrument designed to measure current should have a very low, ideally zero, resistance so that its inclusion in series with the current does not alter the value of that current.

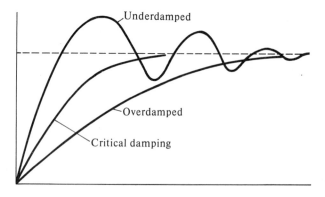

Fig. 10.1.

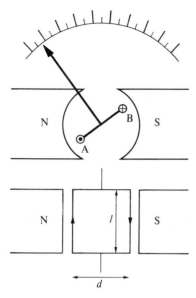

Fig. 10.2.

The moving-coil instrument

Basically the moving-coil instrument consists of a coil in a magnetic field, and depends on the motor principle (Fig. 10.2). Current in the coil in the direction shown (i.e. into the paper at B and out at A) will try to turn the coil clockwise. Of course, with no control device it would always end up at right angles to the lines of force. This type of instrument will have a coil spring as the control device and the coil will turn until the deflecting torque equals the restoring torque produced by the spring. Let us assume that this occurs at an angular deflection θ. If the restoring torque is proportional to θ, as will occur if the spring is linear, then it is equal to $K\theta$, where K is a constant of the spring.

Now the force on each side of the coil will be $BIlN$ newtons, where B = flux density in Wb/m², I = current being measured (in the coil) (A), l = length of coil sides (m), and N = number of turns on the coil. The torque on each coil side = $NBIl\,d/2$ newton metres

total deflecting torque = $NBIld$ newton-metres.

But this must equal the restoring torque, hence

(10.1) $\qquad NBIld = K\theta$

and, as N, B, l, d, and K are constants

(10.2) $\qquad I \propto \theta,$

that is, the deflection is directly proportional to the current being measured, resulting in a linear scale. Note that

(10.3) $$\theta = \frac{BldN}{K} I$$

and $BldN/K$ is the deflection produced by unit current.

Damping, to prevent undue oscillation, is normally provided by winding the coil on an aluminium former. When the coil moves, eddy currents are induced in the former which, by Lenz's Law, flow in a direction to try to prevent the coil from moving. This has the effect of making movement sluggish, that is of damping the movement. It has no effect when the pointer is not moving so that it does not affect the final rest position of the pointer.

It is essential when dealing with any instrument to understand exactly what it is that is being measured. If the moving coil instrument were measuring a current which was varying slowly, it would simply follow the variations, measuring the *instantaneous* value. It would do this, of course, for very low frequency a.c. However, at frequencies above a few hertz, the movement could not follow the variations and would assume an *average* value. The true average of a sine wave is zero, which is what the instrument would read on all but the lowest frequency sine waves. (Note that the value 0.637 times peak is the average of *half* a cycle—see Chapter 6.) The moving coil instrument is thus an average-reading instrument, and will indicate zero on any symmetrical waveform. For a d.c. which is not steady it will read the average value, unless the variations are very slow.

Moving-coil instruments can be extremely accurate, and can be made with sensitivities of $10\,\mu A$ or less. However, they are fairly delicate and the maximum current which they can pass is about 50 mA.

The moving-iron instrument

The moving-iron instrument depends on the principle of attraction or repulsion between magnetic poles. One type, the *attraction* type, consists of a coil which passes the current being measured. This produces a magnetic field which attracts a piece of soft iron, the movement of which moves a pointer across a scale. Most modern instruments are of the *repulsion* type. Two rods of soft iron are situated inside the coil carrying the current and become magnetized with the same polarity. One is fixed, the other being movable. The moving iron is repelled and in moving moves the pointer across the scale.

As the force of repulsion is proportional to the pole strength of each iron, and as the pole strength is proportional to the current in the coil

deflecting torque $\propto$ square of the current.

If the control torque is produced by a spring and is proportional to deflection

(10.4) $\qquad$ deflection $\propto$ (current)^2.

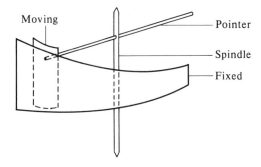

Fig. 10.3.

Of course, if the current is varying it will read the mean of i^2, and hence is a true r.m.s. reading instrument, i.e. it reads d.c. or the r.m.s. value of an a.c. The scale will, of course, be non-linear, although it can be made more linear by suitably shaping the irons. A section of a typical moving iron instrument is shown in Fig. 10.3. Damping is usually provided by a piston moving in a cylinder.

Moving iron instruments are cheaper and more robust than moving coil ones, but are far less sensitive and accurate. The maximum sensitivity is about 10 mA, and, because the coil is fixed, they can be made to pass currents up to about 100 A. Their use is usually limited to frequencies below about 10 kHz. They cannot, therefore, be used with waveforms containing components above this frequency.

Rectifier instruments

Although the moving iron is a true r.m.s. instrument, it suffers some serious defects. As has been explained, moving-iron instruments are not very sensitive or accurate and are limited in their frequency range. The moving-coil instrument can be used on a.c. if the a.c. is rectified. Of course, it is of little value to place a rectifier in series with the instrument, as this would rectify the current being measured. What is used is a bridge rectifier (Fig. 10.4), which allows the current being measured to flow in both directions, as indicated. However, the current in the instrument is full-wave rectified (Fig. 10.5) and the reading will be the average of this. For a sine wave

(10.5) $$I_{av} = 0.637 I_{pk}.$$

Although the rectifier instrument reads the average value of the waveform it is often calibrated in r.m.s. values on the assumption that the waveform is sinusoidal. As the r.m.s. value of a sine wave is 1.11 times the average value (1.11 is the form factor) the readings are, in effect, scaled up by 1.11. Great care must be taken when measuring non-sinusoidal quantities. The only thing that can be said with certainty is that the average value is

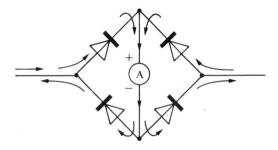

Fig. 10.4.

Fig. 10.5.

the instrument reading divided by 1.11. The r.m.s. value can then be found if, and only if, the form factor is known.

For example, if such an instrument is used on a square-wave current and is reading 10 mA, the r.m.s. value is 10/1.11 or 9.01 mA. This is because a square wave has a form factor of unity.

Shunts

Shunts are used to increase the range of an ammeter. As has been stated earlier, the maximum current which can be passed through a moving coil instrument is in the order of 50 mA. To measure currents larger than this a resistance called a *shunt* is placed in parallel with the instrument to divert part of the current (Fig. 10.6).

Consider, for example, a moving-coil instrument with a full scale deflection (FSD) of 15 mA and a coil of resistance 5 Ω. Let us calculate the value of resistance needed in parallel to enable its FSD to be increased to 10 A. When the instrument is reading full scale it is passing 15 mA and the voltage across the coil must be

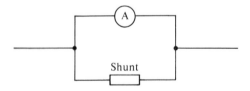

Fig. 10.6.

$$15\,\text{mA} \times 5 = 75\,\text{mV}.$$

If it is to measure 10 A the shunt must pass

$$10 - 0.015 = 9.985\,\text{A}.$$

As it is in parallel with the instrument, it too must have 75 mV across it. Hence the shunt resistance, R_s, is given by

$$R_s = \frac{0.075\,\text{V}}{9.985\,\text{A}} = 0.00751\,\Omega.$$

This, as we might have expected, is a very low value of resistance indeed. Shunts are normally made of maganin strip and may be in the instrument case or externally connected.

10.2 Measurement of voltage

A frequently used method of measuring voltage is to use an ammeter to measure the current produced by the voltage in a given resistor. A perfect voltmeter would, of course, take no current from the voltage being measured, because any current taken would in theory alter the voltage.

A moving-coil ammeter may be connected in series with a resistor, called a *multiplier*, to the voltage being measured. The current flowing will be proportional to the voltage, and the instrument may be scaled in volts. Because the current should be as small as possible, the ammeter should have as high a sensitivity as possible.

Consider the ammeter with a FSD of 15 mA and a resistance of 5 Ω used above. If we wish to use it as a voltmeter with a FSD of 100 V, we connect it as shown in Fig. 10.7. At full scale I must equal 15 mA. As, under these circumstances, the voltage is 100 V, the total resistance must be

$$R = \frac{100\,\text{V}}{15\,\text{mA}} = 6.667\,\text{k}\Omega$$

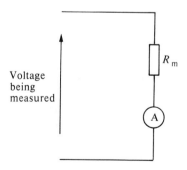

Fig. 10.7.

and as the instrument resistance is 5 Ω

$$R_m = 6667 - 5 = 6662 \, \Omega.$$

This current of 15 mA may be excessive, and considerably alter the voltage being measured as will be seen below. Consider an ammeter with a FSD of 50 μA and a resistance of 5 Ω. Then

$$R = \frac{100 \, \text{V}}{0.05 \, \text{mA}} = 2000 \, \text{k}\Omega \text{ or } 2 \, \text{M}\Omega.$$

R_m will also be, in effect, 2 MΩ (more accurately 1 999 995 Ω!) If this same 50 μA ammeter were used to measure 1 V the total resistance would then be

$$R = \frac{1 \, \text{V}}{0.05 \, \text{mA}} = 20 \, \text{k}\Omega.$$

In fact, it should be clear that any voltage range will have a total resistance of 20 kΩ for each volt of FSD (i.e. 20 kΩ for 1 V FSD, 2 MΩ for 100 V FSD etc.). Thus the current (at FSD) taken by the voltmeter is often quoted in terms of the resistance per volt. The instrument considered above would be said to have a resistance of 20 kΩ per volt, i.e. it takes a full scale current of 1 V/20 kΩ or 50 μA. The higher the 'ohms per volt' the nearer the instrument is to being a perfect voltmeter.

The moving-iron instrument can also be used with a multiplier, although the resistance is usually included by making the coil the required resistance. For example, a moving-iron instrument may be such that 200 ampere-turns are needed on the coil to produce full-scale deflection. If it is required to use it as a 100 V (full-scale) instrument with a current of 10 mA

$$\text{turns on coil} = \frac{200}{0.01} = 20\,000 \text{ turns}$$

$$\text{resistance of coil} = \frac{100 \, \text{V}}{0.01 \, \text{A}} = 10\,000 \, \Omega.$$

Effect of current on voltage being measured

Because the type of voltmeters considered above take current from the voltage being measured, they must have some effect on this voltage. Consider Fig. 10.8, where the voltage at point X is to be measured. If the instrument of resistance 20 kΩ per volt is being used on the 100 V range it will clearly have no noticeable effect when connected between X and earth in parallel with 100 Ω. However, the circuit of Fig. 10.9 is very different. Like Fig. 10.8 the 'true' voltage at X is 50 V. A voltmeter of resistance 20 kΩ/V makes the lower 1 MΩ resistor become 667 kΩ and the voltage at X is

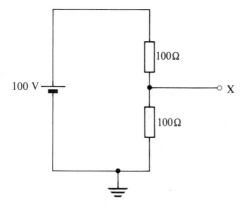

Fig. 10.8.

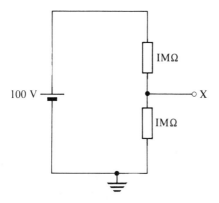

Fig. 10.9.

$$\frac{667 \times 100}{1667} = 40 \text{ V}$$

an error of 20 per cent.

The electrostatic voltmeter

This type of voltmeter takes no current. It consists of a moveable plate which is free to rotate into a fixed hollow plate. The voltage being measured is applied between the fixed and moving plates and the force of attraction moves the moving plate and thus carries a pointer across a scale. Like the moving-iron instrument, it is a true r.m.s. reading instrument. Such voltmeters are very insensitive and are only used when the current taken by other voltmeters would prove excessive.

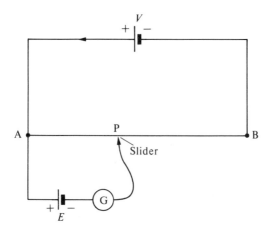

Fig. 10.10.

The d.c. potentiometer

The d.c. potentiometer is a laboratory instrument used for very accurate measurements. Many commercial potentiometers are available. Here we shall only discuss the principles of operation.

A battery, V, is used to supply current to a piece of uniform resistance wire AB (Fig. 10.10). The wire is stretched on a scale so that the distance AP can be measured. As the wire is uniform this length is proportional to the voltage between A and P.

E is, at first, a standard cell, such as a *Weston cadmium cell* which develops 1.0186 V when no current is being taken from it. G is a galvanometer, that is a very sensitive moving-coil instrument. With the slider pressed onto the wire, position P is adjusted until there is no galvanometer deflection, that is no current is flowing through the galvanometer. In this position the voltage between A and P must equal that of the standard cell, E_s. Let the distance AP be l_1.

E_s is now replaced by the voltage to be measured, E, and the new length of AP to cause no deflection is found, say l_2. The ratio of E_s to E must equal the ratio of l_1 to l_2.

(10.6)
$$\frac{E}{E_s} = \frac{l_2}{l_1}.$$

This method of measurement is a so called *null method* because we are not measuring a current, but only detecting the absence of one. The galvanometer need not be accurate, only sensitive. Further, the eye is very sensitive to slight movements of the pointer, so that the null point can be very accurately determined. No energy at all is taken from the voltage which is being measured. Potentiometers can also be used for measuring current (by passing it through a known resistance) and resistance.

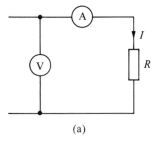

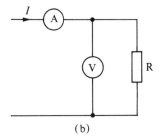

Fig. 10.11.

10.3 Measurement of resistance

The simplest method of measuring resistance is to measure the voltage across and current through the unknown resistor. However, two methods of connection are possible (Fig. 10.11). In Fig. 10.11 (a) the voltmeter is measuring the voltage across the ammeter and the resistor in series. If the resistance of the ammeter is known it can, of course, be subtracted from V/I. In Fig. 10.11 (b) the ammeter measures the current taken by the voltmeter in addition to that taken by the resistor. If the resistance of the voltmeter is R_v,

$$I = \frac{V}{R} + \frac{V}{R_v}.$$

Hence $\dfrac{V}{R} = I - \dfrac{V}{R_v}$

(10.7) or $R = \dfrac{V}{I - \dfrac{V}{R_v}}.$

The *ohmmeter* is a version of this method. It forms part of commercially available *multimeters*, such as the AVO. Such instruments are moving-coil instruments and are switched to measure d.c. currents and voltages using internal shunts and multipliers. They also measure alternating quantities

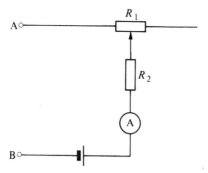

Fig. 10.12.

using a bridge rectifier and are calibrated to read the r.m.s. value of a sine wave.

When switched to one of the resistance ranges a circuit such as Fig. 10.12 is used, A and B being the instrument terminals to which the unknown resistance is connected. Before connecting the unknown, A and B are connected together and R_1 adjusted to give full-scale deflection which is marked zero on the ohms ranges. The unknown is now connected and the instrument reading (lower than full-scale deflection) calibrated in resistance values. Zero current will correspond to an open circuit, or a resistance of infinity.

The Wheatstone bridge

The ammeter/voltmeter method (or the multimeter) is good enough for most practical purposes. However, on some occasions greater accuracy is required and the *Wheatstone bridge* is used. Like the potentiometer, this is a null method of measurement and so does not depend on the accuracy of an instrument. The circuit is shown in Fig. 10.13.

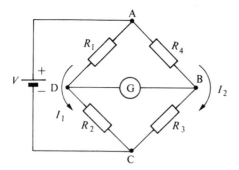

Fig. 10.13.

One resistor, say R_1, is the unknown and the others are adjusted until there is no current in the galvanometer. The bridge is then said to be balanced and the potentials of points B and D must be the same, or, in other words, the voltages across R_2 and R_3 must be equal.

Now $$I_1 = \frac{V}{R_1 + R_2} \quad \text{(no current in DB)}$$

and $$I_2 = \frac{V}{R_3 + R_4}.$$

$$\text{Voltage across } R_2 = \frac{VR_2}{R_1 + R_2}.$$

$$\text{Voltage across } R_3 = \frac{VR_3}{R_3 + R_4}.$$

Hence $$\frac{VR_2}{R_1 + R_2} = \frac{VR_3}{R_3 + R_4}$$

$$R_2 R_3 + R_2 R_4 = R_1 R_3 + R_2 R_3$$

$$R_2 R_4 = R_1 R_3$$

(10.8) or $$R_1 = \frac{R_2 R_4}{R_3}.$$

R_2 and R_3 are usually switched resistors of value 1, 10, 100, and 1000 Ω so that the ratio R_2/R_3 can be set in powers of ten from 1/1000 to 1000/1. R_2 and R_3 are known as the ratio arms and R_4 may be varied between zero and about 10 000 Ω.

As explained earlier, the galvanometer accuracy is not important and neither is the value of the supply voltage, V. Measurements may be made with a very high degree of accuracy.

Examples 10.1

1. A rectifier instrument employing a moving-coil ammeter is calibrated to read the r.m.s. value of a sine wave. When measuring a voltage with a form factor of 1.05 it reads 220 V. Determine the average and r.m.s. values of the voltage.

Solution

$$\text{average value} = \frac{220}{1.11} = 198.2 \text{ V}$$

$$\text{r.m.s. value} = 198.2 \times 1.05 = 208.11 \text{ V}.$$

2. A moving-coil instrument has a full scale deflection of 15 mA and a resistance of 5 Ω. How could it be used to give a full-scale deflection of (a) 5 A and (b) 250 V?

Solution

(a) Meter voltage (FS) = 15 mA × 5 = 75 mV.

Shunt current = 5 − 0.015 = 4.985 A.

$$R_s = \frac{0.075}{4.985} = 0.015\,\Omega \text{ connected in parallel.}$$

(b) Total resistance = $\frac{250\,V}{15\,mA}$ = 16.667 kΩ.

R_m = 16 667 − 5 = 16 662 Ω in series.

3. The shunt of question 2 is too made of manganin with a resistivity of 50 μΩ cm. If the manganin is 0.5 mm thick and 10 cm long find the width required.

Solution

$$R_s = 0.015 = \frac{\rho l}{a}$$

$$a = \frac{\rho l}{R_s} = \frac{50 \times 10^{-6} \times 10}{0.015}$$
$$= 0.0333\,cm^2.$$

Width = $\frac{0.0333}{0.5 \times 10^{-1}}$ = 0.67 cm.

4. A voltmeter of resistance 10 kΩ/V is connected, on the 50 V range, to measure the voltage between X and Y (Fig. 10.14). Calculate the true and measured values of this voltage.

Solution

Voltmeter resistance = 500 kΩ.

500 kΩ in parallel with 400 kΩ = 222.2 kΩ.

1 MΩ in parallel with 1 MΩ = 500 kΩ.

$$V_{XY} = \frac{500 \times 100}{972.2} = 51.4\,V \text{ (true value).}$$

The voltmeter in parallel with 1 MΩ in parallel with 1 MΩ gives 250 kΩ.

Measured V_{XY} = $\frac{250 \times 100}{722.2}$ = 34.6 V.

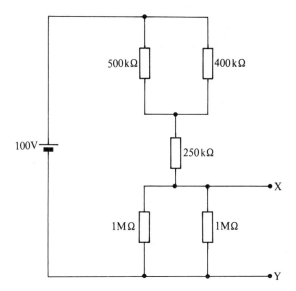

Fig. 10.14.

10A Test questions

1. State the three forces or torques present in a deflection instrument.

2. The control torque
    (a) converts the quantity being measured into a torque,
    (b) prevents undue oscillation,
    (c) is used to make the scale more linear,
    (d) opposes the deflecting torque.

3. The damping torque
    (a) converts the quantity being measured into a torque,
    (b) prevents undue oscillation,
    (c) is used to make the scale more linear,
    (d) opposes the deflecting torque.

4. State two methods of producing the damping torque.

5. An ammeter should ideally have a resistance
    (a) which is very high,
    (b) which is very low,
    (c) which is zero,
    (d) equal to the circuit resistance.

6. What would be the effect, in a moving-coil instrument, of having no
    (a) control device,
    (b) damping?

7. How are damping and control torques provided in most moving-coil instruments?

8. A moving-coil instrument has a 50-turn coil of area 4 cm². The control spring constant is 10^{-6} Nw-m per degree deflection and the flux density is 0.2 Wb/m². Find the deflection produced by a current of 1 mA.

9. A moving-coil instrument has a coil of dimensions 2 cm by 3 cm. The control spring constant is 0.8×10^{-6} Nw-m per degree and the flux density is 0.25 Wb/m². If it is to give a full scale deflection of 65° with a current of 10 mA, find the number of turns required on the coil.

10. The moving-coil instrument measures
    (a) instantaneous values,
    (b) r.m.s. values,
    (c) peak values,
    (d) average values.
(More than one answer is possible.)

11. A moving-coil instrument is measuring a sinusoidal current of peak value 1 mA. It will read
    (a) 1 mA,
    (b) 0.637 mA,
    (c) 0.707 mA,
    (d) zero.

12. The instrument of question 11 is included in a bridge rectifier circuit to measure the same current. The reading will be
    (a) 1 mA,
    (b) 0.637 mA,
    (c) 0.707 mA,
    (d) zero.

13. A voltage of 1 V peak is connected to a moving-coil instrument in series with a 1 kΩ resistor and a perfect diode. The instrument will read
    (a) zero,
    (b) 0.3185 mA,
    (c) 0.637 mA,
    (d) 0.707 mA.

14. If, in question 12, the instrument were calibrated to read the r.m.s. value of a sine wave, it would read
    (a) 1 mA,
    (b) 0.637 mA,
    (c) 0.707 mA,
    (d) zero.

15. A voltage of form factor 1.08 is being measured by a rectifier instrument calibrated to read the r.m.s. value of a sine wave. If it reads 150 V, determine the average and r.m.s. values of the voltage.

16. Describe the action of a repulsion type moving-iron instrument, including a description of the methods of control and damping. Give the relationship between current and deflection and explain how the linearity of the scale can be improved. What are the main advantages and disadvantages compared to a moving-coil instrument?

17. Draw a circuit diagram of a bridge rectifier instrument, showing clearly the current flow on each half cycle.

18. A moving-coil ammeter has a full scale deflection of 50 μA and a resistance of 100 Ω. Show how it can be used as
    (a) an ammeter with FSD of 100 mA,
    (b) a voltmeter with FSD of 100 V.
In each case show how any external components are connected.

19. A moving-coil instrument has a FSD of 15 mA and a resistance of 5 Ω. If a shunt of resistance 0.02 Ω is connected in parallel with it, determine the current which will produce full-scale deflection.

20. A moving-coil instrument has a flux density of 0.1 Wb/m^2 and a coil of 60 turns of area 10 cm^2. If the control spring constant is 10^{-6} Nw-m per degree, and the FSD is 50°, find the series resistance needed to use the instrument as a voltmeter with a full-scale deflection of 500 V. The resistance of the coil may be neglected.

21. A multi-range meter is quoted as being '10 kΩ per volt' on the d.c. voltage ranges. What is the resistance on the 100 V range?

22. A moving-iron instrument requires 150 ampere-turns for FSD. Find the number of turns and resistance of the coil if it is to be used as
    (a) an ammeter of FSD 10 A
    (b) a voltmeter of FSD 250 V and taking a current of 20 mA at full scale.

23. The electrostatic voltmeter is particularly useful for measuring small voltages
    (a) true,
    (b) false.

24. The electrostatic voltmeter measures
    (a) average values,
    (b) r.m.s. values,
    (c) peak values.

25. Discuss the principles of 'null' methods of measurement.

26. A potentiometer is used to measure the e.m.f., E, of a cell. The positions for balance for the cell E and a standard cell of e.m.f. 1.0186 V are 23.8 cm and 46.73 cm respectively. Find the e.m.f. of the cell E.

27. The value of a resistor R is being measured by the circuit of Fig. 10.11 (a). If the voltmeter has a resistance of 500 Ω/V and the ammeter a resistance of 5 Ω, determine R if the readings are 2.5 V (on the 10 V range) and 20 mA respectively.

28. If the same instruments are now used on the same ranges in the circuit of Fig. 10.11 (b), find their readings. Assume that the supply being used is not changed.

29. What percentage error would be obtained in questions 27 and 28 if the meter resistances had been neglected?

30. Draw a circuit diagram of a Wheatstone bridge and derive the balance equation.

31. In the circuit of Fig. 10.13, R_1 is the unknown resistance. If R_2 is 10 Ω and R_3 is 100 Ω find the value of R_1 if R_4 is 72.5 Ω at balance.

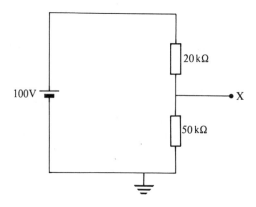

Fig. 10.15.

32. If the voltage source V and the galvanometer of Fig. 10.13 are interchanged, will the balance equation be altered?

33. A voltmeter of resistance 500 Ω/V is used on the 100 V range to measure the voltage between X and earth in the circuit of Fig. 10.15. Determine the percentage error in the reading.

34. If the voltmeter of question 33 is now used on its 50 V range, find the reading which it would give.

35. A voltmeter of FSD 10 V and resistance 500 Ω/V is connected across the 2.5 kΩ resistor in the circuit of Fig. 10.16. It reads 5 V. Determine
   (a) the value of R,
   (b) the 'true' voltage across the 2.5 kΩ resistor (i.e. without the meter),
   (c) the 'true' voltage across R,
   (d) the reading which the voltmeter would give if connected across R.

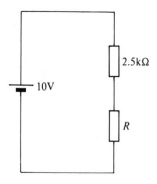

Fig. 10.16.

10.4 Measurement of power

The electrodynamic instrument

The *electrodynamic instrument* (*dynamometer*) is a type of moving-coil instrument. The magnetic field in which the moving coil moves is produced by current flowing in a pair of stationary coils. Whilst such an instrument can be used as an ammeter (and hence as a voltmeter) by passing the current to be measured through the fixed and moving coils in series or parallel, it is most frequently used as a *wattmeter*, to measure *power*.

To measure the power in a load, the current through the load is passed through the fixed coils, and the voltage across the load causes a current in the moving coil, which has a resistance in series with it (Fig. 10.17). Notice the markings M, L, V+, and V which are used on the instrument. It can be shown that the deflection is proportional to the mean power, or, to be more exact, for a sinusoidal waveform, $V I_L \cos\phi$ where ϕ is the phase angle between V and I_L.

A slight error is introduced because the power in the load (which is what we are trying to measure) is not $V I_L \cos\phi$ but $V_L I_L \cos\phi$. The error is, of course, due to the power being dissipated in the fixed coils and is equal to $I_L^2 R_f$ so that

(10.9) $$\text{power in load} = \text{wattmeter reading} - I_L^2 R_f.$$

If I_L is very large, leading to serious errors, the method of connection shown in Fig. 10.18 may be used. In this case an error due to the power dissipated in the moving coil is introduced and

(10.10) $$\text{power in load} = \text{wattmeter reading} - \frac{V_L^2}{R_m}.$$

To summarize: the connection of Fig. 10.17 is the better for high-voltage low-current situations; that of Fig. 10.18 for low-voltages and high currents.

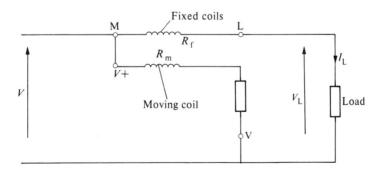

Fig. 10.17.

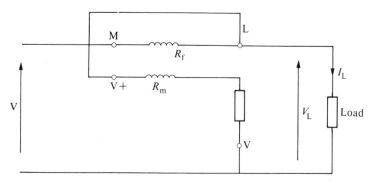

Fig. 10.18.

10.5 The cathode-ray oscilloscope as a measuring instrument

The *cathode-ray oscilloscope* (CRO) is an instrument normally used to display a graph of the voltage to which it is connected, against time. It will be described in TEC units on electronics.

Whilst it is of extreme value in observing the shape of waveforms, its use as a measuring instrument must not be overlooked. The vertical deflection of the beam will be calibrated in volts/cm, and might range from a few millivolts to a hundred or so volts/cm. Most CROs have a d.c. input so that the distance which a d.c. voltage moves the spot can be used to measure the voltage. As far as a.c. is concerned, the peak or peak-to-peak values of a waveform are easily obtained, whatever the shape of the waveform.

The horizontal deflection is calibrated in time/cm, so that the period, and hence the frequency, of the waveform can be measured. For both voltage and time measurements, the student should study the handbook of the particular oscilloscope being used. These measurements must be practised in he laboratory.

The CRO is also useful for measuring the phase difference between two waveforms. In this case the time base is not used, the two voltages being applied to the horizontal (X) and vertical (Y) plates respectively. If they are in phase they will produce a straight line on the screen at 45° to the horizontal if both inputs are adjusted to give the same peak-to-peak deflections. Similarly, if they are 90° out of phase the spot traces a circle on the screen. In general, a trace such as that shown in Fig. 10.19 is produed and it can be shown that if the phase angle between the waveforms is ϕ

(10.11) $$\sin \phi = \frac{x}{y}.$$

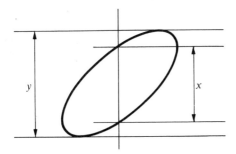

Fig. 10.19.

10.6 Conclusion

This chapter has presented a brief outline of various measuring instruments and various methods of measurement. Little has been said about electronic and digital instruments. Their method of operation is best left for an electronics text.

Electronic voltmeters amplify the voltage being measured and can be very sensitive. They have very high input resistances, usually in the order of $10\,M\Omega$ and are thus much closer to being 'true' voltmeters than moving-coil instruments with multipliers. They can easily be used to measure current, resistance, or power.

Many instruments in use today give a digital display similar to that on a calculator or digital watch. The first digital instruments on the market were counters used to measure frequency or time intervals. These are relatively simple, just counting the number of cycles in a given time interval. The digital voltmeter has now been available for some years and is a very accurate and sensitive instrument. There is very little chance of error due to the user, and the digital information available can, if needed, be supplied directly to a computer or control system. Many firms now manufacture digital multimeters, measuring a.c. and d.c. voltage and current as well as resistance. With prices continually falling, these instruments will no doubt shortly replace the conventional types.

10B Test questions

1. The dynamometer wattmeter will not measure the power in a d.c. circuit,
    (a) true,
    (b) false.

2. Draw circuit diagrams showing the two most frequently used methods of connecting a dynamometer wattmeter to measure the power in a load, indicating when each method would be the more suitable.

3. A dynamometer wattmeter has fixed coils connected in series of total resistance $0.5\,\Omega$ and a moving coil of $2.5\,\Omega$ in series with a $100\,\Omega$ resistor. A resistive load takes a current of 2 A when connected directly to a constant 10 V supply. Determine the power in the load and the reading given by the wattmeter when connected
   (a) as Fig. 10.17,
   (b) as Fig. 10.18 to the 10 V supply.
Note: the addition of the wattmeter might alter the load voltage and/or current.

4. Discuss the use of the CRO as a measuring instrument.

5. Explain how the CRO may be used to measure the phase difference between two sinusoidal voltages of the same frequency.

Solutions to test questions

1A
1. 2 A. 2. 6 V. 3. 2 kΩ. 4. 50 V.
5. 20 W, 18 W, 2 mW, and 250 μW respectively.
6. (a) 50 Ω.
   (b) 2 A in each.
   (c) 40 V and 60 V respectively.
   (d) 2 A.
   (e) 80 W and 120 W respectively.
   (f) 200 W.
7. (a) 12 Ω.
   (b) 5 A and 3.33 A respectively.
   (c) 100 V.
   (d) 8.33 A.
   (e) 500 W and 333 W respectively.
   (f) 833 W.
8. (a) 6.55 V.
   (b) 6.55 mA, 3.28 mA, and 2.18 mA respectively.
   (c) 42.9 mW, 21.48 mW, and 14.28 mW respectively.
   (d) 78.7 mW.
9. 28.8 Ω. 10. 18 Ω. 11. 1.4 mm. 12. 0.73 Ω.
13. −0.00042. 14. 3.71 Ω. 15. 149°C. 16. 0.05 Ω.
17. (a) 2.1 V.
    (b) 1.62 W and 1.08 W respectively.
    (c) 0.45 W.

2A
1. 1.75 S 3.5 A. 2. 1 A, 0.5 A, and 2 A respectively.
3. 24 A total. 4A, 20 A, 6 V. 4. 0.4 A downwards.

2B
1. 0.4 A downwards.
2. 3.6 A upwards in the 6 V battery and 3.2 A downwards in the 4 V battery.
3. 0.22 A left to right.

2C
1. 0.22 A left to right. 2. 0.278 A. 3. 7.5 Ω, 0.052 W.
4. 1 A. 5. 0.1 A right to left.

3A
1. a. 2. b. 3. a. 4. b. 5. b. 6. a. 7. c. 8. a. 9. a. 10. d.
11. a. 12. 0.15 C. 13. 112.5 J. 14. 50 nF; 37.5 μC.
15. 7.5 μC; 11.25 μC; 18.75 μC.
16. (a) 1 μF. (b) 200 μC. (c) 100 V; 66.7 V; 33.3 V.
    (d) 0.01 J; 0.0067 J; 0.0033 J.

226 *Solutions to test questions*

17. 6.37 nF.   18. 2.1 nF.   19. 30 kV.
20. 40 V. Note: C is five times as large so that V must be divided by 5 to keep $Q (= CV)$ constant.

4A 1. a.   2. b (not *exactly* at the geographic north pole).   3. b.   4. c.
5. b.   6. c.   7. b.   8. 0.09 Wb/m^2.   9. 200 N.   10. 333 A.   11. a.

4B 1. 50 V.   2. 4000 V.   3. 502.7 V.   4. 0.05 mWb.   5. 100 m/s.
6. 0.2 Wb/m^2.

4C 1. a.   2. a.   3. a.   4. 1313.   5. 1.5 mWb.   6. 0.82 mWb.
7. 1.89 A.   8. 4.88 A.

4D 1. 4.4 mH.   2. 177.   3. 22 V.   4. $0.88 \, \mu\text{Wb}$.   5. $1.38 \, \mu\text{J}$.

5A 1. a.   2. a.   3. b (they all do).   4. a.   5. b.   6. a.   7. b.   8. b.
9. d.   10. 3.67 A.
11. (a) 0.25 s.   (b) 2.43 A.   (c) zero.
12. (a) 0.01 s.   (b) 0.5 A.   (c) 0.43 A.
(d) 50 A/s. At initial rate it would reach 0.5 A in 0.01 s.
13. 1.3 A, 0.87 A.
14.

time	0	0.25	0.5	0.75	1.00	1.25	1.50	1.75	2.00
ch.	0	4.42	7.87	10.55	12.64	14.27	15.54	16.52	17.29
disch.	20	12.13	7.36	4.46	2.71	1.64	1.00	0.60	0.37

15. (a) 200 s.   (b) 174.7 V.   (c) $2.5 \, \mu\text{A}$.   (d) $0.75 \, \mu\text{A}$.
16. (a) 1 s.   (b) $10 \, \mu\text{A}$.   (c) $10 \, \mu\text{A/s}$.   (d) 10 V/s.   (e) 3.93 V.
(f) $6.07 \, \mu\text{A}$.
17. (a) 0.1 mA.   (b) $22.3 \, \mu\text{A}$.   (c) 11.15 V.
18. (a) 10 V.   (b) 3.9 V.

6A 1. b and f.   2. b.   3. (a) 141.4 V.   (b) −61.8 V.
4. (a) 166 V. (Note: 200 V is the r.m.s. value.)   (b) zero.
5. 1.4 ms.   6. Av. = 127.4 V. r.m.s. = 141.4 V.   7. 1.08.
8. 407 V lagging V_1 by 25°.   9. 407 V lagging V_1 by 25°.
10. $72.1 \sin(\omega t + 0.442)$.   11. 907 V leading v_3 by 26'.   12. Zero.

6B 1. $62.83 \text{ k}\Omega$ $15.92 \, \mu\text{A}$ lagging by 90°.   2. 238.7 kHz.
3. (a) $161 \, \Omega$.   (b) 62.1 mA.   (c) 51.5° lagging.   (d) 0.39 W.
4. $22.2 \, \Omega$ 0.079 H.
5. (a) $10 \, \Omega$.   (b) 0.154 H.   (c) 4.86 A.   (d) 78.3° lagging.
(e) 236 W.
6. $7.1 \, \mu\text{F}$.
7. (a) $557 \, \Omega$.   (b) 3.59 mA leading by 51°.   (c) 4.5 mW.
(d) 2.26 V.   (e) 3.8 V.

Solutions to test questions 227

6C 1. c. 2. a. 3. d. 4. b.
5. (a) 796 Hz. (b) 20 mA. (c) 0.18 W. (d) 4V.

7A 1. a. 2. c. 3. 866 W. 4. 4.16 A.
5. (a) 7.27 A lagging by 72.3°. (b) 0.3. (c) 1745 VA.
  (d) 1665 VAr. (e) 528 W.
6. (a) 159 μF. (b) 50 V. (c) 60 W. (d) 0.6.
7. 16 Ω, 0.095 H.
8. (a) 11.3 A lagging by 41.2°. (b) 2041 W. (c) 1786 VAr.
  (d) 2712 VA.

7B 1. a. 2. 1.84 MHz, 17.3. 3. 5.63 mH, 53. 4. 123.5 pF, 1.43 MHz.
5. 36.5 Hz, 1.4. 6. d.

7C 1. (a) inductor 637 μA
    capacitor 943 μA
    total 306 μA.
  (b) I leads V by 90°. (c) 4.11 MHz. (d) 775 μA.
  (e) infinity (total current zero).
2. 0.17 μF; 2.5 W; 65.4° current leading. 3. 1.75 μF or 8.38 μF.
4. (a) coil 0.67 mA; capacitor 1.18 mA. (b) 0.71 mA; 59.5° lead.
  (c) 3.52 kΩ (d) 0.9 mW.

7D 1. 2.6 MHz.
2. (a) 3.33 kΩ. (b) 0.75 mA in phase.
  (c) coil 0.97 mA; capacitor 0.61 mA. (d) 1.9 mW.
3. (a) 129 kHz. (b) coil 122.4 mA; capacitor 121.6 mA.
  (c) 15 mA. (d) 6.7 kΩ. (e) 8.2.
4. (a) resistor 4.8 A; inductor 5.1 A.
  (b) 7 A lagging by 46.7°. (c) 1.15 kW.
5. (a) 237 μF. (b) 97.6 Hz. 6. 154.5 μF. 7. 2 nF.
8. 2340 μF; 184 kVA. 9. 1702 μF; 103.4 kVA; 413.2 A; 252.8 A.

7E 1. (a) 4.3 A. (b) 2785 W. (c) 0.85.
2. (a) line 10.46 A; phase 6.04 A. (b) 5470 W.
3. 70.4 A. 4. 7.14 A; 3200 W.

8A 1. (a) 48 V. (b) 0.096 A; 0.48 A. (c) 0.52 A; 2.58 A.
  (d) 23 W; 66.6 W.
2. (a) 417. (b) 0.12 V/turn. (c) 1.25 A. (d) 0.26 A.
  (e) 923 Ω.
3. 1.8×10^{-14} W; 35.4:1 step-down; 5.63×10^{-12} W.
4. (a) 3200 Ω. (b) 50 V; 0.25 A. (c) 200 V; 0.0625 A.
  (d) 12.5 W.

228 *Solutions to test questions*

5. (a) 98.6 per cent.   (b) 98 per cent.
6. 5.29 A at power factor 0.983.   7. 5.79 A at power factor 0.726.
8. (a) 25.3 kW.   (b) 96.2 per cent.

9A
1. See text (p.187).   2. See text (p.185).
3. (a) $Z/2p$.   (b) $Z/2$.   4. (a) 2000 V.   (b) 1000 V.   5. 985 V.
6. (a) 542 V.   (b) 400 V.   7. 85.7 mWb.   8. 427.5 V.

9B
1. (a) 242.5 V.   (b) 277 rev/min.   (c) 209 Nm.
2. 774 rev/min.   3. 1328 rev/min.   4. 991 rev/min.
5. 1311 rev/min.   6. 0.29 Ω.

9C
1. 523.4 rev/min.   2. 1.24 Ω.   3. 39.8 Ω.

10A
1. See text (p.204).   2. d.   3. b.   4. See text (p.206).   5. c.
6. See text (p.205).   7. See text (p.205).   8. 4°.   9. 34.7.
10. a, d.   11. d.   12. b.   13. b.   14. c.
15. 135.14 V; 145.9 V.   16. See text (p.206).
17. See text (p.208).
18. (a) 0.05 Ω in parallel.   (b) 1 999 900 Ω in series.
19. 3.765 A.   20. 60 kΩ.   21. 1 MΩ.
22. (a) 15 turns, as low a resistance as possible.
    (b) 7500 turns, 12.5 kΩ.
23. b.   24. b.   25. See text (p.212).   26. 0.519 V.
27. 120 Ω.   28. 20.46 mA; 2.4 V.
29. +4.2 per cent; −18.7 per cent.   30. See text (p.214).
31. 7.25 Ω.   32. No.   33. −22.2 per cent.   34. 45.5 V.
35. (a) 1.67 kΩ.   (b) 6 V.   (c) 4 V.   (d) 3.33 V.

10B
1. b.   2. See text (p.221).
3. (a) 16.5 W; 18.2 W.   (b) 16.4 W; 17.2 W.
4. See text (p.222).   5. See text (p.222).

Index

a.c. 90
    circuits 106
        capacitance in 110
        inductance in 107
        resistance in 106
acceptor circuits 154
active component 127
alternating quantity 90
ampere 1
    definition of 33, 51
angular velocity 92
armature 187
average value 93
    of square wave 94

back e.m.f. 191
bridge, Wheatstone 214
brushes 185

capacitance 35
capacitor 36
    air 43
    ceramic 43
    charging 84
    discharging 86
    electrolytic 43
    energy stored in 39
    mica 43
    paper 43
    parallel plate 38
capacitors in parallel 37
    series 37
cathode-ray-oscilloscope 222
ceramic capacitor 43
charge 33
chemical effect of current 4
coercive force 70
coercivity 70
commutator 185
conductance 8
control torque 204
corkscrew rule 48
coulomb 53
CR circuit (a.c.) 112
    (d.c.) 84
current 1
    chemical effect of 4
    heating effect of 4
    magnetic effect of 4, 47
    magnification 154
    measurement of 204
cycle 90

damping torque 204
d.c. generator 185
    motor 191
        losses 202
        series 199
        shunt 193
        starter 202
    potentiometer 212
deflecting torque 204
delta connection 167
dielectric 36
    constant 39
    strength 40
digital instrument 223
dynamic impedance 152
dynamometer 221

eddy currents 179
efficiency, d.c. motor 203
    transformer 180
electric field 33
    flux 33
electrolytic capacitor 43
electromagnetic induction 54
electron 33
electrostatic voltmeter 211
e.m.f. 1
energy stored in capacitor 39
    inductor 73
exponential curve 77

farad 36
Faraday 54
field current 189
field, electric 33
    magnetic 46
Fleming's left hand rule 49
    right hand rule 56
flux density 50
flux, electric 33
    magnetic 46
force on conductor 49
form factor 94
Franklin 33
frequency 90
fringing 64

heating effect of current 4
Henry 71
Hertz 90
hysteresis 70
    loop 71

impedance 113
    dynamic 152
    triangle 116
inductance 71

inductance (cont.)
   electromagnetic 54
   mutual 73
   self 71
inductor 71
instantaneous value 92
internal resistance 5
iron losses 179

Kirchhoff's 1st law 17
   2nd law 18

lap winding 188
LCR circuit (a.c.) 117
leakage 64
left hand rule, Flemming's 49
Lenz's law 54
line voltage 164
linear component 1
losses, motor 202
   transformer 179
LR circuit (a.c.) 112
   (d.c.) 76

magnet 46
magnetic circuits 58
   effect of current 4
   field 46
      strength 59
   flux 46
      density 50
   poles 46
magnetising current 171
   force 59
magneto-motive-force 58
magnification, voltage 124, 132
   current 154
matching 175
maximum power theorem 23
measurement 204
mica capacitor 43
m.m.f. 58
moving coil instrument 205
moving iron instrument 206
multimeter 213
multiples 5
multiplier 209
mutual inductance 73

neutral 164
non-linear component 1
null method 212

Oersted 47
ohmmeter 213
Ohm's law 1
open circuit test 180

paper capacitor 43
parallel circuit 1

(a.c.) 136
   place capacitor 38
   resonance 150
peak to peak value 92
peak value 92
period 90
periodic time 90
permeability 59
   of free space 59
   relative 60
permittivity 39
   of free space 39
   relative 39
phase 95
   measurement of 222
   voltage 164
phasor 98
   diagram 98
poles, magnetic 46
polyphase 162
potential 34
   difference 1, 34
   gradient 35
potentiometer 212
power 4
   factor 126
      correction 155
   in a.c. circuits 126
   in capacitor 111
   in inductor 109
   maximum (theorem) 23
   measurement of 221
primary 170
principle of superposition 12

Q factor 124, 131, 154

reactance, capacitive 111
   inductive 109
reactive component 127
rectifier instrument 207
rejector circuit 154
relative permeability 60
   permittivity 39
reluctance 59, 62
remanence 70
remanent flux density 70
resistance 1
   internal 5
   measurement of 213
resistivity 3
resonant frequency 122, 152
resonance, parallel 150
   series 122
right hand rule, Flemmings 56
right hand screw rule 48
r.m.s. value 94

secondary 170
self inductance 71

separately excited generator 189
series circuit 1
    (a.c.) 112
series motor 199
short circuit test 180
shunt motor 193
shunts 208
sine wave 91
slip rings 185
solenoid 48
star connection, four wire 164
    three wire 167
starter 202
submiltiples 5
superposition 12

temperature coefficient 3
terminal voltage 5, 189
Thévenin's theorem 24
three phase supply 162
three wire star 167
time constant 78, 85

torque 192
transient 76
transformer 71, 170
    losses (copper) 180
    (iron) 179
    open circuit test 180
    short circuit test 180
triangle, impedance 116

VA 126
VAr 128
volt 1
voltage 1
    measurement 209
voltmeter, electrostatic 211

wattmeter 221
wave winding 188
weber 50
Wheatstone bridge 214

yoke 187

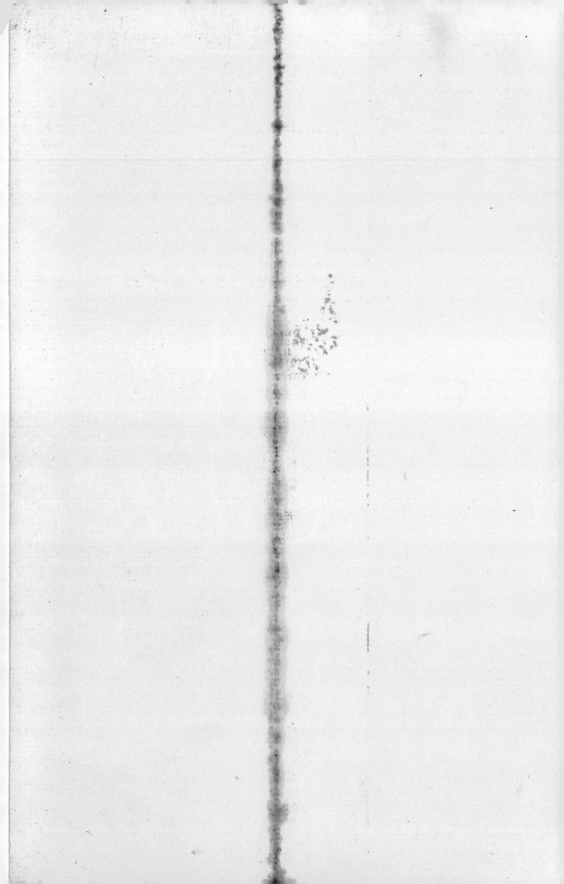